Behavioral
Cybersecurity

Behavioral Cybersecurity

Fundamental Principles and Applications of Personality Psychology

Wayne Patterson and Cynthia E. Winston-Proctor

CRC Press

Taylor & Francis Group

Boca Raton London New York

CRC Press is an imprint of the
Taylor & Francis Group, an **informa** business

First edition published 2021
by CRC Press
6000 Broken Sound Parkway NW, Suite 300, Boca Raton, FL 33487-2742

and by CRC Press
2 Park Square, Milton Park, Abingdon, Oxon, OX14 4RN

© 2021 Taylor & Francis Group, LLC

CRC Press is an imprint of Taylor & Francis Group, LLC

Library of Congress Cataloging-in-Publication Data
Names: Patterson, Wayne, 1945- author. | Winston-Proctor,
Cynthia E., author.
Title: Behavioral cybersecurity. Fundamental principles and applications of
personality psychology / Wayne Patterson, Cynthia Winston-Proctor.
Description: First edition. | Boca Raton : CRC Press, 2021. | Includes
bibliographical references and index.
Identifiers: LCCN 2020027781 (print) | LCCN 2020027782 (ebook) |
ISBN 9780367509798 (hardback) | ISBN 9781003052029 (ebook)
Subjects: LCSH: Computer security. | Computer fraud. | Hacking. |
Self-protective behavior.
Classification: LCC QA76.9.A25 P38455 2021 (print) | LCC QA76.9.A25
(ebook) | DDC 005.8—dc23
LC record available at https://lccn.loc.gov/2020027781
LC ebook record available at https://lccn.loc.gov/2020027782

ISBN: 9780367509798 (hbk)
ISBN: 9781003052029 (ebk)

Typeset in Times
by codeMantra

CONTENTS

Authors   xi

Introduction   xiii

1   Recent Events   1
     Addressing DDoS Attacks   4
     Ransomware   5
     Facebook "This is Your Digital Life"   7
     Yu Pingan   8
     The (US) Department of Justice Success in Prosecuting
     Cybercriminals: Who's Winning?   8
     "Fake News" Concerning the Coronavirus   10
     Problems   10
     References   11

2   Behavioral Cybersecurity   12
     Cybersecurity Without the Human: Is It Only a Matter
     of Time?   13
     Cybersecurity and Personality Psychology: Why This
     Field of Psychological Science?   13
     References   14

3   Personality Theory and Methods of Assessment   16
     Personality Traits and the "Social Actor"   17
     Personality Characteristic Adaptations and "the
     Motivated Agent"   19
          Extrinsic Motivation versus Intrinsic Motivation   20
          Power Motivation: Striving for Power   20
          Social Motivation: The Striving for Affiliation   21

Achievement Motivation: The Striving for
Achievement   21
The Narrative Identity Dimension of Human
Personality: The Autobiographical Author   22
Conclusion   24
References   25

4   Hacker Case Studies: Personality Analysis and Ethical
Hacking   28
Comrade   28
Adrian Lamo   29
Gabriel   29
Hacker Personality Descriptions   30
Ethical Hacking   31
Programs to Encourage the Development of Ethical
Hackers   32
Problems   33
References   34

5   Profiling   35
Profiling in the Cybersecurity Context   35
Sony Pictures Hack   36
Profiling Matrices   37
The "ABCD" Analysis   40
Problems   44
Reference   44

6   Access Control   45
Authentication   45
Something You Know: Passwords   45
Good Password Choice   46
Password Meters   47
Tokens: What You Have   49
Biometrics: What You Are   49
Problems   50
Reference   51

7   The First Step: Authorization   52
Security Levels   52
Partial and Total Order   53
Covert Channel   54
Inference Control   55
Inference Control and Research   55
A Naïve Answer to Inference Control   56
Randomization   56

	Firewalls	56
	Problems	56
	References	58
8	Origins of Cryptography	59
	Caesar Shift	59
	Substitution and Transposition	60
	The Keyword Mixed Alphabet Cryptosystem	61
	The Vigenère Cryptosystem	62
	One-Time Pad Encryption	62
	The Playfair Square	64
	Rotor Machines	66
	World War II and the Enigma Machine	66
	Problems	66
	References	68
9	Game Theory	69
	Payoff	69
	Matrix Games	72
	Mixed Strategy	72
	Saddle Points	73
	Solution of All 2 × 2 Games	74
	Dominated Strategies	74
	Graphical Solutions: 2 × n and m × 2 Games	75
	Using Game Theory to Choose a Strategy in the Sony/North Korea Case	77
	References	79
10	The Psychology of Gender	80
	Definitions and Analysis of Gender	81
	Gender-As-Trait: The Sex Differences Approach	81
	Gender in Social Context: The Within Gender Variability Approach	81
	Gender Linked to Power Relations Approach	82
	Gender as Intersectional: The Identity Role, Social Identity, and Social Structural Approach	82
	The Nature versus Nurture Debate in Gender Psychology	83
	Conclusion	83
	References	84
11	Turing Tests	86
	Introduction	87
	The Role of the Turing Test in Behavioral Cybersecurity	88
	A Final Exam Question	88

	While Grading	89
	Turing's Paper in Mind	89
	The Imitation Game	90
	Respondents	91
	Summary of Results	92
	"Coaching" Respondents	93
	Future Research	95
	Problems	95
	References	96
12	Modular Arithmetic and Other Computational Methods	97
	Z_n or Arithmetic Modulo n	98
	What are the Differences in the Tables?	100
	Finite Fields	101
	The Main Result Concerning Galois Fields	102
	Matrix Algebra or Linear Algebra	102
	Problems	103
13	Modern Cryptography	106
	Modern Cryptographic Techniques	106
	The Advanced Encryption Standard	107
	SubBytes	109
	ShiftRow	110
	MixColumns	110
	AddRoundKey	110
	Test Vectors	111
	Symmetric Encryption or Public Key Cryptology	118
	The PKC Model for Key Management	118
	Can We Devise a PKC?	119
	The RSA Public Key Cryptosystem	120
	What Is the RSA Cryptosystem?	120
	Problems	121
	References	122
14	Steganography and Relation to Crypto	123
	A History of Steganography	123
	Transmission Issues	125
	Image Steganography	126
	Image File Formats	126
	An Example	127
	Using Cryptography and Steganography in Tandem or in Sequence	128
	Comments	130
	Problems	130
	References	131

15	A Metric to Assess Cyberattacks	132
	Defining a Cybersecurity Metric	132
	The Attacker/Defender Scenario	132
	Rivest–Shamir–Adleman: An Interesting Example	133
	Attack/Defense Scenarios	134
	Conclusion	136
	Problems	136
	References	136
16	Behavioral Economics	138
	Origins of Behavioral Economics	138
	Utility	139
	Challenge to Utility Theory	140
	Application of the Kahneman–Tversky Approach to Cybersecurity	144
	Nudge	145
	An Application of Nudge Theory to Cybersecurity	146
	Problems	146
	References	146
17	Fake News	147
	A Fake News History	147
	Fake News Resurgence, Acceleration, and Elections	148
	What Is Fake News?	148
	Satire or Fake News?	149
	Distinguishing Satire from Fake News	150
	DailyBuzzLive.com	150
	ABCnews.com.co	151
	TheOnion.com	151
	Infowars.com	151
	New Yorker	152
	Empirenews.net	152
	Beforeitsnews.com	152
	Centers for Disease Control	153
	Assessing Fake (or Not-Fake) News	154
	Problems	156
	References	156
18	Exercises: Hack Labs	157
	Hack Lab 1: Social Engineering: Find Cookie's Password	157
	Hack Lab 2: Assigned Passwords in the Clear	158
	Hack Lab 3: Sweeney Privacy Study	159
	Hack Lab 4: Password Meters	160
	Problems	161
	References	163

19   Conclusions   164
        Profiling   164
        Social Engineering   164
        Sweeney Privacy   165
        Understanding Hackers   165
        Game Theory Application to Profiling   165
        Turing Tests   166
        Crypto and Stego   166
        Behavioral Economics   166
        Fake News   166
        Password Meters   167
        Next Steps   167

Index   169

AUTHORS

Dr. Wayne Patterson is a retired professor of computer science from Howard University. He is also currently coprincipal investigator for the NSF-funded GEAR UP project at Howard, which has supported almost 300 STEM undergrads to do summer research in 15 developing countries. He has also been Director of the Cybersecurity Research Center, Associate Vice Provost for Research, and Senior Fellow for Research and International Affairs in the Graduate School at Howard. He has also been Professeur d'Informatique at the Université de Moncton, Chair of the Department of Computer Science at the University of New Orleans, and in 1988 Associate Vice Chancellor for Research there. In 1993, he was appointed Vice President for Research and Professional and Community Services, and Dean of the Graduate School at the College of Charleston, South Carolina. His other service to the graduate community in the United States has included being elected to the presidency of the Conference of Southern Graduate Schools and also to the Board of Directors of the Council of Graduate Schools. Dr. Patterson has published more than 50 scholarly articles primarily related to cybersecurity. He has been the principal investigator on over 35 external grants valued at over $6,000,000.

He holds degrees from the University of Toronto and the University of New Brunswick and the PhD in mathematics from the University of Michigan. He has also been a Program Manager for the National Science Foundation and a Visiting Scholar at Google.

He has also held postdoctoral appointments at Princeton University and the University of California, Berkeley.

Dr. Cynthia E. Winston-Proctor is a widely respected and accomplished narrative personality psychologist and academic. She is Professor of Psychology and Principal Investigator of the Identity and Success Research Laboratory at Howard University. She is also founder of Winston Synergy LLC, a psychology and education consulting firm. Dr. Winston-Proctor earned her Bachelor of Science degree in psychology from Howard University and her PhD in psychology and education from the University of Michigan. Recognized as an outstanding psychologist, research scientist, and teacher, Dr. Winston-Proctor was awarded the National Science Foundation Early Career Award for scientists and engineers, the Howard University Syllabus of the Year Award, the Howard University Emerging Scholar Award, and a Brown University Howard Hughes Medical Institute Research Professorship. Also, she was elected as a member of the Society of Personology, the oldest and most prominent society for scholars to develop, preserve, and promote theory and research that focuses on the study of individual lives and whole persons. Her theory and method development-focused research and education scholarship have resulted in publications in numerous journals and edited books, including *Culture & Psychology*; *Qualitative Psychology*; *Journal of Research on Adolescence*; *Psych Critiques*; *New Directions in Childhood & Development*; *The Oxford Handbook of Cultural Psychology*; and *Culture, Learning, & Technology: Research and Practice*. Dr. Winston-Proctor's professional service includes serving as an editor on the Editorial Board of the *American Psychological Association Journal of Qualitative Psychology*, President of the Society of STEM Women of Color, Member of the Board of Directors of the Alfred Harcourt Foundation, and Advisor to the Board of Directors of the Howard University Middle School of Mathematics and Science.

INTRODUCTION

Since the introduction and proliferation of the Internet, problems involved with maintaining cybersecurity have grown exponentially and evolved into many forms of exploitation.

Yet, cybersecurity has had far too little study and research. Virtually all of the research that has taken place in cybersecurity over many years has been done by those with computer science, electrical engineering, and mathematics backgrounds.

However, many cybersecurity researchers have come to realize that to gain a full understanding of how to protect a cyberenvironment requires the knowledge of not only those researchers in computer science, engineering, and mathematics but also those who have a deeper understanding of human behavior: researchers with expertise in the various branches of behavioral science, such as psychology, behavioral economics, and other aspects of brain science.

The authors, one a computer scientist and the other a psychologist, have attempted over the past several years to understand the contributions that each approach to cybersecurity problems can benefit from in this integrated approach that we have tended to call "behavioral cybersecurity."

The authors believe that the research and curriculum approaches developed from this integrated approach provide a book with this approach to cybersecurity. This book incorporates traditional technical computational and analytic approaches to cybersecurity, and also psychological and human factors approaches.

The history of science seems to evolve in one of two directions. At times, interest in one area of study grows to the extent that it grows into its own discipline. Physics and chemistry could be described in that fashion, evolving from "natural science." There are other occasions, however, when the underlying approach of one discipline is complemented by a different tradition in a totally separate discipline. The study of computer science can be fairly described as an example of that approach. At the University of Michigan in the 1970s, there was no department of computer science. It was soon born as a fusion of mathematics and electrical engineering.

Our decision to create this book arose from a similar perspective. Our training is in computer science and psychology, and we have observed, as have many other scholars interested in cybersecurity, that the problems we try to study in cybersecurity require not only most of the approaches in computer science but also more and more an understanding of motivation, personality, and other behavioral approaches in order to understand cyberattacks and create cyberdefenses.

As with any new approaches to solving problems when they require knowledge and practice from distinct research fields, there are few people with knowledge of the widely separate disciplines, so it requires an opportunity for persons interested in either field to gain some knowledge of the other. We have attempted to provide such a bridge in this book that we have entitled *Behavioral Cybersecurity*.

In this book, we have tried to provide an introductory approach in both psychology and cybersecurity, and as we have tried to address some of these key problem areas, we have also introduced topics from other related fields such as criminal justice, game theory, mathematics, and behavioral economics.

We entered the computer era almost 75 years ago. For close to two-thirds of that time, we could largely ignore the threats that we now refer to as cyberattacks. There were many reasons for this. There was considerable research done going back to the 1970s about approaches to penetrate computer environments, but there were several other factors that prevented the widespread development of cyberattacks. Thus, the scholarship into the defense (and attack) of computing environments remained of interest to a relatively small number of researchers.

As we will see in the book, beginning in the 1980s, a number of new factors came into play. The world of computer science was thus faced with the dilemma of having to adapt to changing levels of expertise in a very short period of time. The first author of this book began his own research in 1980, in the infancy of what we now call cybersecurity, even before the widespread development of the personal computer and the Internet.

In the attempt to try to address the need for an accelerated development of researchers who can address the problems of cyberattacks, our two authors have recognized that in addition to the traditional expertise required in studying such problems—that is, expertise in computer science, mathematics, and engineering—we also have a great need to address the human behavior, in the first place, of persons involved in cyberattacks or cybercrime of many forms, but also in the behavioral aspects of all computer users, for example, those who would never avoid precautions in their life such as locking their doors, but use the name of their significant other, sibling, or pet as a password on their computer accounts.

As a result, we have embarked on this project in order to introduce into the field an approach to cybersecurity that relies upon not only the mathematical, computing, and engineering approaches but also a greater understanding of human behavior. We have chosen to call this subject area "behavioral cybersecurity" and have developed and offered a curriculum over the past several years that now has evolved into this textbook, which we hope will serve as a guidepost for universities, government, industry, and others that wish to develop scholarship in this area.

Behavioral Cybersecurity provides a basis for new approaches to understanding problems in one of our most important areas of research—an approach, agreed upon by most cybersecurity experts, of incorporating not only traditional technical computational and analytic approaches to cybersecurity but also developing psychological and human factor approaches to these problems. We wish to allow nonspecialists to gain an understanding in their everyday computing environments as to the challenges in avoiding or deterring malicious attackers.

In addition, noting that while only 20% of computer science majors in the United States are women, about 80% of psychology majors are women. It is hoped that this new curriculum, with a behavioral science orientation in the now-popular field of cybersecurity, will induce more women to want to choose this curricular option.

In terms of employment needs in cybersecurity, estimates indicate "more than 209,000 cybersecurity jobs in the United States are unfilled, and postings are up 74% over the past 5 years."

It is believed that the concentration in behavioral cybersecurity will also attract more women students since national statistics show that whereas women are outnumbered by men by approximately 4 to 1 in computer science, almost the reverse is true in psychology.

Our objective with this textbook is to encourage many more opportunities to study and research the area of cybersecurity through this approach to behavioral cybersecurity. With a new approach to the skill set needed for cybersecurity employment, it is hoped that an expanded pool of students will seek to follow this path.

It has also not escaped our notice that the field of cybersecurity has been less attractive to women.

RECENT EVENTS

The history of research in cybersecurity dates back to the 1970s and even before, but for most people, what was known at the time affected only a very small number of people in the world of computing. As has been described earlier, the first general awareness of external attacks occurred only in the 1980s with, for example, the Morris Internet worm of 1988.

On November 2 of that year, Robert Morris, then a graduate student in computer science at Cornell, created the worm in question and launched it on the Internet. In UNIX systems of the time, applications sendmail and finger had weaknesses that allowed the worm to enter and then generate copies of itself. This resulted in the major damage caused by this work, as it would make copies of itself until memory was exhausted, causing the system to shut down. Furthermore, with UNIX vulnerabilities, the worm could move from one machine to the other, and it was estimated that it infected eventually about 2000 computers within 15 hours. The US Government Accountability Office estimated the cost of the damage in the range of $100,000–$10 million—obviously not a very accurate assessment.

Robert Morris was tried and convicted under the Computer Fraud and Abuse Act and was sentenced to 3 years' probation and 400 hours of community service and fined $10,050.

He was subsequently hired as a professor of computer science at the Massachusetts Institute of Technology, where he continues to teach and research to this day.

The proliferation of access to computers to the general public, in both North America and worldwide, only began in the 1990s. Consequently, to most of us, hearing of cyberattacks only began in that period.

But now the percentage of the global population with access to the Internet has increased exponentially, and consequently, not only the number of targets for cyberattackers and undoubtedly the number of attackers have increased, but also the potential for exploitation of many different types of target has increased. Therefore, we now have many examples of exploits that result in widespread theft of critical user information such as Social Security numbers, addresses, telephone numbers, and even credit card information.

The Office of Personnel Management (OPM) is a little-known, but important component of the US government. It is responsible for "recruiting, retaining and honoring the world-class force to serve the American people" (Office of Personnel Management, 2018).

Two major data breaches at the OPM occurred in 2014 and 2015 that obtained the records of approximately 4 million US government employees, with the estimate of the number of stolen records as approximately 21.5 million. Information obtained included Social Security numbers, names, dates and places of birth, and addresses. Also stolen in this attack were 5.6 million sets of fingerprints. As a consequence of these two attacks, both the director and the chief information officer of OPM resigned.

Attacks on many organizations through techniques such as distributed denial of service (DDoS) designed primarily to cripple an organization's website, thus preventing the organization from doing business, even if only for a relatively short period of time.

Perhaps the first time there was a widespread public awareness of DDoS attacks occurred around November 28, 2010. The website wikileaks.org published several thousand classified US government documents and announced they had 250,000 more. This caused an immediate "cyberwar," with opponents of WikiLeaks attempting to crash the WikiLeaks website, whereas WikiLeaks supporters, particularly a loosely affiliated group called Anonymous, were able to crash numerous sites, including MasterCard and PayPal.

In 2007, *WikiLeaks* was founded as an organization (WikiLeaks, 2018) and a website whose mission was to:

> bring important news and information to the public
>   provide an innovative, secure and anonymous way for sources to leak information to journalists, and
>   publish original source material alongside our news stories so readers and historians alike can see evidence of the truth.

Alas, WikiLeaks released several thousand classified US government documents and announced they had 250,000 more. (It was later determined that the documents had been provided by US Army Private Bradley/Chelsea Manning.)

The immediate response from the cyber community was that two of the main sources of receipt of donations to WikiLeaks, MasterCard and PayPal, announced that they would no longer accept donations destined as contributions to WikiLeaks, its primary source of revenue.

Subsequently, organizations supportive of the objectives of WikiLeaks, notably an online group called Anonymous, decided to launch DDoS attacks on both MasterCard and PayPal and were successful in bringing them down for several days, undoubtedly costing both those organizations significant loss of revenue. In retaliation, in effect perhaps the first example of cyber warfare, other groups sympathetic to MasterCard and PayPal and in opposition to WikiLeaks release of the classified documents made a similar attempt to take down the WikiLeaks website (see Figure 1.1).

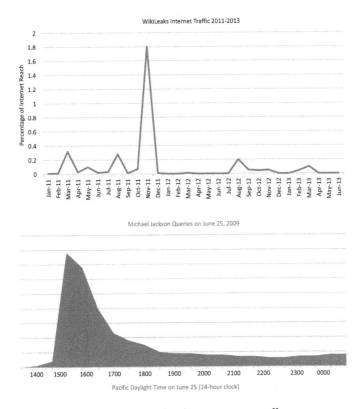

Figure 1.1 Two measures of spikes in Internet traffic, suggesting a possible DDoS attack, or just heightened public interest. (a) WikiLeaks website showing attack level after Manning documents leaked. (Data from Alexa.com.); (b) Internet Traffic Report on June 25, 2009 (Michael Jackson's death). (Data source: Wired.com.)

ADDRESSING DDoS ATTACKS

DDoS attacks are coordinated efforts by human or machine to overwhelm websites and, at a minimum, to cause them to shut down. The use of this type of malicious software has grown exponentially in the past decade, and despite considerable research, it has proven very difficult to identify, detect, or prevent such attacks. On the other hand, increases in traffic at websites may not be the result of a DDoS attack but a legitimate increase in demand for the Web service.

In the 2011 publication of the World Infrastructure Security Report (WISC, 2011), it was noted that the reported increase in DDoS attacks had been multiplied by a factor of 10 since the first year of the study in 2004 and that ideologically motivated "hacktivism" and vandalism have become the most readily identified DDoS attack motivations, and since 2014, the number and intensity of DDoS attacks has been increased exponentially.

In its simplest form, a DDoS attack is a coordinated set of requests for services, such as a Web page access. These requests may come from many nodes on the Internet, by either human or electronic action, and the requests require resource utilization by the site under attack. The Low Orbit Ion Cannon (LOIC) is easily accessible on the Internet, and use of this software to initiate or participate in a DDoS attack only requires typing in the website name.

A DDoS attack might be indistinguishable from a sudden influx of requests because of a specific event. For example, news sites might have an extraordinary increase in legitimate requests when a significant event occurs—the death of a celebrity, for example, or a ticket brokerage service may be flooded when popular tickets go on sale.

There may be numerous reasons for unusual Web traffic: There may be a cycle in the business of the host site, for example, stock prices at the moment of the opening bell in the stock market, at university home pages on the last day of course registration, or with the "Michael Jackson phenomenon"—when Michael Jackson died, most news sites reported a heavy spike in their Web traffic because of the widespread curiosity in users attempting to discover what had occurred, or there may be an actual DDoS attack underway.

The original reporting on these data came as a result of a joint research team consisting of students and faculty from Howard University in Washington, DC, and the Universidad Santo Tomás in Santiago, Chile, working together to develop this research (Banks et al., 2012).

RANSOMWARE

Ransomware surfaced in 2013 with CryptoLocker, which used Bitcoin to collect ransom money. In December of that year, ZDNet estimated based on Bitcoin transaction information that the operators of Crypto Locker had procured about $27 million from infected users.

Phishing attacks are a form of illicit software designed primarily to obtain information to benefit "Phisher" from an unsuspecting person or account.

These attacks might arise from any source the user contacts for information, and many might occur from opening an email supposedly from some trusted source.

The purpose for the attack might be to urge a recipient to open an attachment. Many users might not realize that opening a Word, Excel, and PowerPoint document may contain within it code (called a macro) that may then infect the system.

Another approach in the phishing attack might be to encourage the recipient to follow a link which purports to require the user to enter an account name, password, or other personal information, which then is being transmitted to the creator of the phishing attack. In such a case, the personal information transmitted may be used by the phishing perpetrator in order to gain other resources of the victim (Figure 1.2).

One example from 2016 involves an email allegedly from the PayPal Corporation. Here is a screenshot of the attack itself. In this case, the objective of the attack is to fool the recipient into believing that this is a legitimate email from PayPal, asking for "help resolving an issue with your PayPal account?" and consequently passing on the login information to the phishing attacker.

A prudent user would look carefully at the email address paypal@outlook.com and realize that it was a bogus email address.

The year 2010 produced a "game changer." For perhaps the first time, a malicious hardware and software attack, coined as Stuxnet, infected nuclear facilities in Iran. One critical difference here was that previous malware always was produced by individuals or small groups, sought random targets, easily disabled when identified, and caused relatively minimal damage.

Stuxnet was discovered by security researchers. It was determined to be a highly sophisticated worm that spread via Windows and targeted Siemens software and equipment. Different versions of

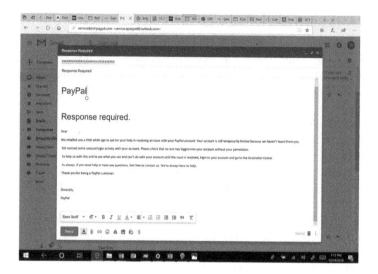

Figure 1.2 A phishing attack using PayPal (facsimile).

Stuxnet infected five Iranian organizations, presumably related to the uranium enrichment infrastructure.

The Iranian nuclear program was damaged by Stuxnet, as the infected control system created a change in temperature in the core, thus destroying the equipment. Kaspersky Labs concluded that the attacks "could only have been conducted with nation-state support." Later reports indicated that Stuxnet was a joint effort of the US National Security Agency and the Israeli comparable agency, the Mossad.

In what might be better known to readers, we have in 2017 and 2018 examples from the *2016 US Presidential Election.*

The use of a private email server by Hillary Rodham Clinton while she was secretary of state during the presidential administration of Barack Obama has sparked a reoccurring debate about the use of private email servers by US government officials. In 2016, Hillary Rodham Clinton was the first woman to win the Democratic Party's nomination for the president of the United States. As is the case in many political situations, there are some angles of the issue that can be more fully explored to understand the nature of the issue. An interesting question to explore the landscape of gender psychology and behavioral cybersecurity is as follows: How might different common conceptualizations of gender along with corresponding

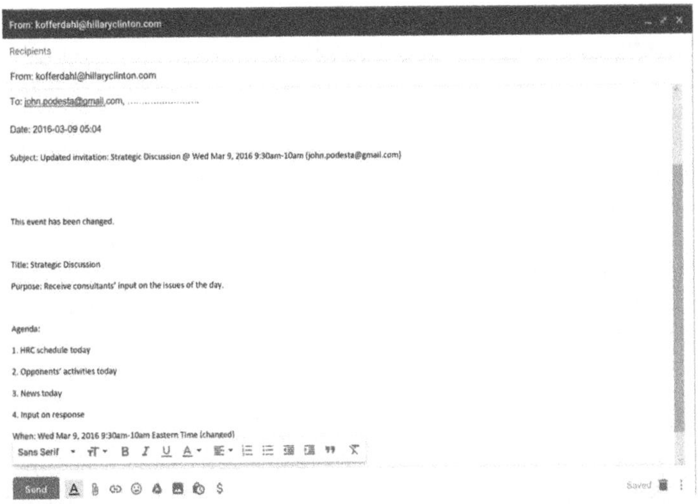

Figure 1.3   Podesta email hacked and sent to WikiLeaks (facsimile).

approaches to understanding the psychology of gender explain the behavior of Secretary of State Hillary Clinton and what caused her use of a private email server?

Attacks that succeeded in obtaining significant email traffic from the internal communications of the US Democratic Party and its presidential candidate, Hillary Clinton, in all likelihood having a significant impact on the outcome of the election. In particular, Clinton's campaign chairman John Podesta had many of his emails stolen, as he was tricked by an outsider's phishing attack (Figure 1.3).

Early in 2018 were ransomware attacks known as WannaCry and Petya (WikiPedia, 2017a, b); the former apparently put close to 100,000 computers up for ransom, including computing systems in many hospitals in the United Kingdom, whereas Petya disabled many industries in Ukraine.

FACEBOOK "THIS IS YOUR DIGITAL LIFE"

The personal data of approximately 87 million Facebook users were acquired via the 270,000 Facebook users who used a Facebook app called "This Is Your Digital Life." As a result of users giving this third-party app permission to collect their data, they were also giving the app access to information on using their friends' data. Underlying this test was a personality assessment of the common Big Five personality traits. This case revealed multiple aspects of interesting questions about personality tests, methods, and assessment. For example, how can an individual's personality traits be measured through a social media application in a way that yields valid and reliable personality data that can be applied to impact advertising strategies and politics? In a hypothetical world, how can the possible motivational dynamics of the researchers, businesspeople, and social media company be assessed using personality test, methods, and assessment?

YU PINGAN

In August 2017, a sealed indictment was filed, in the Southern District of California, against the Chinese national named Yu Pingan (also known as Gold Sign, on a Charge of Conspiracy Computer Hacking. (US, 2018)

The charge alleges that Mr. Pingan provided malicious software, in particular a software tool known as "Sakula," to four companies, two in California, one in Massachusetts, and one in Arizona. Then this software was capable of downloading to victims computers without authorization. Secular was then alleged to have been necessary for large-scale breaches including the OPM attack as well as the health insurer Anthem.

Mr. Pingan was at the time a 36-year-old resident of Shanghai, China, but he was subsequently arrested because he flew to the United States to attend a conference.

THE (US) DEPARTMENT OF JUSTICE SUCCESS IN PROSECUTING CYBERCRIMINALS: WHO'S WINNING?

In order to gain an understanding of the relative success of cybercriminals as compared with law enforcement and its ability to deter cybercriminals, for many years the data regarding cyberattacks were very poorly understood. In some ways, this continues to be the case, but there is now a slightly clearer picture, certainly because to some extent there is a more coordinated effort to report on prosecutions for what is considered computer crime.

One major reason for this is that since the beginning of the computer era and the corresponding security concerns with the appearance of viruses, worms, denial of service, and ransomware, the law has simply not involved as rapidly as the technology. For example, consider the following: I may have a very important file or set of files that can realistically be assessed as having a significant monetary value. For the sake of discussion, let us say that this information is worth $1 million. Now, a cybercriminal somehow successfully copies all of this information. Has a theft occurred?

Of course, our instinct would say yes. But if you look at any legal definition of theft, this action cannot be described this way, since the original owner still possesses that information. It is apparent that there is conflict in the definition of the term. On the one hand, obtaining electronic information without the permission of the owner satisfies part of the definition of theft, but on the other hand, the legitimate owner still retains the same information.

A number of years ago, the US Department of Justice began to categorize press releases related to legal actions, usually successful prosecutions, related to a category called "Cyber Crime." These releases can be found at the website: https://www.justice.gov/news?f%5B0%5D=field_pr_topic%3A3911.

It should also be noted, appropriately, that the DoJ only reports crimes that fall under the jurisdiction of the U.S. federal government and not similar crimes that might be violations of state or local law. Nevertheless, aggregating these data would allow us to project an average approximately 30 such cases per year; in particular, 21 in 2015, 37 in 2016, 23 in 2017, and 29 in 2018.

But Sophos reports that "the total global number of malicious apps has risen steadily in the last four years". In 2013, just over ½ million samples were malicious. By 2015 it had risen to just under 2.5 million. For 2017, the number is up to nearly 3.5 million. The vast majority are truly malicious with 77% of the submitted samples turning out to be malware.

This is a worldwide sample; however, the U.S. data represent 17.2% of the global figure. Thus, we could estimate reasonably that for 2017, the prevalence of grants and more attacks in the United States would exceed 500,000.

It would lead one to believe that the odds are pretty good for the cybercriminal.

"FAKE NEWS" CONCERNING THE CORONAVIRUS

The worldwide pandemic COVID-19, or the coronavirus, has sparked its own plethora of fake news meant to confuse the public and serve to the advantage—for political, monetary, or other malicious reasons—of the propagators of these items of fake news.

A few of many examples include statements by the President of Brazil pandemic as "only a mild flu." Both Twitter and Facebook have deleted posts by him as fake news. The President of the United States suggested in a news conference that COVID-19 could be eradicated by drinking bleach. The Mafia in Italy has distributed information about getting (for a substantial fee) a virus test, which is fake in itself. One county official in Florida said that the coronavirus can be killed by holding a blow dryer up to your nose. One post originating in England offered a 5000 pound credit to "help people through the Corona virus prices" if they can collect 520 pounds in cash; someone would come around to pick it up. Also in England, a text in Facebook claims the British Army has been called in to help with the response to the virus—but the military images are taken out of context. Another tweet claims the Dutch Air Force was going to disinfect the entire country with helicopters. And finally, among many others, a screenshot on Twitter claimed that Russia had unleashed 500 lions to keep people indoors.

PROBLEMS

1. Estimate the number of computers infected by the Morris worm.
2. Discuss the GAO estimates of the cost of the Morris worm. Find alternative estimates.
3. What organization or organizations have given awards to WikiLeaks for humanitarian efforts?
4. Find an organization for which one can make donations through PayPal.
5. Identify a ransomware attack and an estimate of the amount of funds paid to the perpetrator or perpetrators.
6. What is the difference between phishing and spear phishing?
7. How many emails related to the 2016 Presidential Election were estimated to be released by WikiLeaks?
8. You can find on the WikiLeaks website the text of the emails to and from John Podesta during the 2016 Presidential

Election campaign. Select ten of these for further analysis. Rank them in terms of their potential damage.

9. In the Department of Justice news website (https://www.justice.gov/news?f%5B0%5D=field_pr_topic%3A3911), find the number of cases in 2017 and 2018 involving:
   a. Hacking
   b. Phishing
   c. DDoS
   d. Ransomware

Is each on the increase, decline, or nonexistent (according to DoJ)?

REFERENCES

Alexa Internet, Inc. 2018. Alexa. https://www.alexa.com.

Banks, K. et al. 2012. DDoS and other anomalous web traffic behavior in selected countries. *Proceedings of IEEE SoutheastCon 2012*, March 15–18, 2012, Orlando, FL.

Department of Justice. 2018. Office of Public Affairs, Russian National Charged with Interfering in U.S. Political System, October 19, 2018. https://www.justice.gov/opa/pr/russian-national-charged-interfering-us-political-system.

Office of Personnel Management. 2018. OPM. https://www.opm.gov.

US. 2018. U.S. v. PINGAN YU Case No. 17CR2869-BTM. United States of America, Plaintiff, v. PINGAN YU, aka "GoldSun," Defendant. United States District Court, S.D. California, August 9, 2018.

WikiLeaks. 2018. WikiLeaks. https://wilileaks.org.

WikiPedia. 2017a. "WikiPedia, Petya Ransomware Attack," *WikiPedia*, May 2017. https://en.wikipedia.org/wiki/Petya_(malware).

WikiPedia. 2017b. "WikiPedia, WannaCry Ransomware Attack." *WikiPedia*, May 2017. https://en.wikipedia.org/wiki/WannaCry_ransomware_attack.

WISC. 2011. World Infrastructure Security Report 2010. Arbor Networks, Chelmsford, MA.

2

BEHAVIORAL CYBERSECURITY

More than ever before in history, the tasks that compose everyday life crisscross digital spaces. On the one hand, this modern reality had led individuals and organizations to celebrate the innovation and sophistication of computer networks and information systems that drive their digital lives. This innovation and sophistication have enabled quicker and more efficient manufacturing, production, critical infrastructure, and supply chain management, as well as access to things such as online shopping, education, banking, and education. At the same time on the other hand, this new digital landscape has opened up a world of tremendous vulnerability for the cybersecurity of both individuals and organizations.

With the explosion of cyberattacks in recent years, the importance of cybersecurity has grown almost without bound. In order to gain an understanding of how to combat these threats, it is necessary to understand cybersecurity from a number of reference points. First, it is imperative to understand the *approaches* available to design a healthy defense strategy. Second, a necessary component to understand the role of defense is being able to identify the possible attack *strategies*. Finally, often omitted in cybersecurity research on the technological approaches and strategies is how they can be compromised by *human behavior*. Thus, we have developed a behavioral cybersecurity model for research, development of higher education curricula, and the pursuit of safer digital lives for individuals and organizations.

This book builds upon our first, as well as the more recent, work we have conducted with our colleagues to explore both the technological and human behavioral issues that are integral to understand cybersecurity. Within the past 6 years, we have led and worked with our colleagues to introduce a new scientific field, behavioral cybersecurity. As we have done so, we have developed the following technical definition of the field: Behavioral cybersecurity integrates behavioral science theories, methods, and research findings to answer questions about behavior involved in the identification, management (i.e., defenses, collateral damage control), and analysis of cybersecurity events faced by individuals, as well as within organizations (Patterson et al., 2016a,b; Patterson & Sone, 2017, 2018; Patterson et al., 2017; Patterson & Winston-Proctor, 2019; Patterson, Murray et al., 2020; Patterson & Gergely, 2020; Patterson, Orgah, et al., 2020). The purpose of this chapter is to introduce the reader to the new field of behavioral cybersecurity.

CYBERSECURITY WITHOUT THE HUMAN: IS IT ONLY A MATTER OF TIME?

Can you imagine the nature and consequences of one of the recent cybersecurity events that we described in Chapter 1, where the human actor is removed? One might argue that the time will come when the human is out of the loop.

Understanding human behavior is integral to cybersecurity. Without a human actor, virtually all cybersecurity issues would be nonexistent. Within computer science and engineering, human factor psychology is the most common psychological subfield used to solve problems. Human factor psychology is a scientific discipline that studies how people interact with machines and technology to guide the design of products, systems, and devices that are used every day, most often focusing on performance and safety (Bannon, 1991; Bannon & Bodker, 1991; Salvendy, 2012). Sometimes, human factor psychology is referred to as ergonomics or human engineering. Within our initial formulation of our behavioral cybersecurity model, we extend this human factor psychology focus to include personality psychology.

CYBERSECURITY AND PERSONALITY PSYCHOLOGY: WHY THIS FIELD OF PSYCHOLOGICAL SCIENCE?

Personality psychologists study the whole person with respect to the following three dimensions of personality that develop within the complex sociocultural context in which individuals' lives develop: personality traits, personality characteristic adaptations, and narrative identity. We will explore each of these dimensions of personhood and how personality psychologists assess each within Chapter 3. We started the development of our behavioral cybersecurity model with personality psychology because it provides a robust framework to describe human nature (i.e., how all individuals are alike), individual differences (i.e., how some individuals are alike), and human uniqueness (i.e., how an individual in some ways is like no other person). Personality psychology is also the only subfield of psychological science that at its founding in the 1930s had its distinctive mission to understand human uniqueness, which is an integral to gain a full understanding of behavioral cybersecurity approaches, strategies, and behaviors.

In sum, with the interest and rapid acceleration of cyberattacks worldwide, it has become clear to us that it is insufficient to gain a comprehensive understanding of the overall landscape of cybersecurity for us to only explore mathematics- and engineering-related approaches within cybersecurity research, education, and public information dissemination. The expertise of personality psychologists along with their colleagues from their sister disciplines of cognitive, developmental, neuro, health, and developmental psychology is imperative. Thus, behavioral cybersecurity is a necessary new discipline with potential to transform the way in which individuals and organizations conceive of cybersecurity and respond within the new normal of modern digital life.

REFERENCES

Bannon, L. J. 1991. From human factors to human actors: The role of psychology and human-computer interaction studies in system design. In J. Greenbaum & M. Kyng (Eds.), *Design at Work: Cooperative Design of Computer Systems.* L. Erlbaum Associates, Hillsdale, NJ, pp. 25–44.

Bannon, L. J., & Bodker, S. 1991. Beyond the interface: Encountering artifacts in use. In J. M. Carroll (Ed.), *Designing interaction: Psychology at the Human Computer Interface.* Cambridge University Press, New York, pp. 227–253.

Patterson, W., Boboye, J., Hall, S., & Hornbuckle, H. 2017. The gender Turing test. *Proceedings of the 3rd International Conference on Human Factors in Cybersecurity,* July 2017, Los Angeles, CA.

Patterson, W., & Gergely, M. 2020. Economic prospect theory applied to cybersecurity. *Proceedings of the AHFE 2020 International Conference on Human Factors in Cybersecurity,* July 16–20, 2020, San Diego, CA, pp. 113–121.

Patterson, W., Murray, A., & Fleming, L. 2020. Distinguishing a human or machine cyberattacker. *Proceedings of the 3rd Annual Conference on Intelligent Human Systems Integration,* Modena, Italy, February 2020, pp. 335–340.

Patterson, W., Orgah, A., Chakraborty, S., & Winston-Proctor, C. E. 2020. The impact of Fake News on the African-American community. *Proceedings of the AHFE 2020 International Conference on Human Factors in Cybersecurity*, July 16–20, 2020, San Diego, CA, pp. 30–37.

Patterson, W., & Sone, M. 2017. Behavioural cybersecurity: A new metric to assess cyberattacks. *Proceedings of Contemporary Mathematics and the Real World*, University of Ibadan, May 2017, Nigeria.

Patterson, W., & Sone, M. 2018. A metric to assess cyberattacks. *Proceedings of the 4th International Conference on Human Factors in Cybersecurity*, July 2018, Orlando, FL.

Patterson, W., & Winston-Proctor, C. E. 2019. An international extension of Sweeney's Data Privacy Research. Advances in Human Factors in Cybersecurity, T. Ahram & W. Karwowski (Eds.), *Proceedings of the AHFE 2019 International Conference on Human Factors in Cybersecurity*, July 24–28, 2019, Washington, DC, pp. 28–37.

Patterson, W., Winston-Proctor, C. E., & Fleming, L. 2016a. Behavioral cybersecurity: Human factors in the cybersecurity curriculum. *Proceedings of the 2nd International Conference on Human Factors in Cybersecurity*, July 2016, Orlando, FL.

Patterson, W., Winston-Proctor, C. E., & Fleming, L. 2016b. Behavioral cybersecurity: A needed aspect of the security curriculum. *Proceedings of the IEEE SoutheastCon 2016*, March 2016, Norfolk, VA.

Salvendy, G. 2012 (Ed.). *Handbook of Human Factors and Ergonomics* (4th ed.). Wiley and Sons, Hoboken, NJ.

3

PERSONALITY THEORY AND METHODS OF ASSESSMENT

What do we know when we know a person? There are many ways to answer this question. Personality psychologists' answer to this question is that to know a person means that you understand the stable dimensions of their social behavior (i.e., personality traits), the adaptive aspects of their needs, wants, motives, and goals (i.e., personality characteristic adaptations), and the person's internalized, evolving, and integrative inner narrative (i.e., narrative identity). These dimensions of knowing a person constitute a person's personality. The purpose of this chapter is to briefly describe the dimensions of human personality and methods personality psychologists use to assess each. We adopt this focus because at the heart of our foundational development of the field of behavioral cybersecurity has been to integrate cybersecurity with a personality psychological approach to behavioral science. From this perspective, we have begun to develop a field of behavioral cybersecurity that centers on understanding the attacker's personality in terms of how it can inform questions about behavior involved in the identification, analysis, and management of cybersecurity events faced by individuals, as well as within organizations (Patterson & Winston-Proctor, 2019).

In his book entitled *The Art and Science of Personality Development*, Dan McAdams (2015) presents the heuristic to understand the personality of a person by thinking of the person as intricately behaving across the life course as a social actor, the motivated agent, and the author. This is his shorthand way to identify and describe each of the three layers of human personality that psychologists have discovered over more than a century of research: personality traits, personality characteristic adaptations, and narrative identity. Each of these dimensions of personality has a differential relation to culture (McAdams & Pals, 2006). To further explain McAdams' (2015) heuristic for conceptualizing personality, he described that across the course of a person's life, we know an actor who first has a style of presentation or temperament that gradually morphs into personality traits; an agent with a dynamic arrangement of evolving motives, goals, values, and personal projects; and an author seeking to understand who they are and how they are to live, love, and work within the social and cultural context of their adult society. Thus, applications of personality psychology to identify, analyze, and manage cybersecurity problems require an understanding of cross-cultural variations in personality traits, motivation, and narrative identity within the social contexts in which cybersecurity problems emerge within and across cultures.

PERSONALITY TRAITS AND THE "SOCIAL ACTOR"

Personality traits are a person's broad, cross-situationally consistent, and relatively enduring basic behavioral tendencies. Some personality scholars have characterized personality traits as an element of personality that another person could interpret not knowing a person very well. Sometimes, psychologists think of this dimension as "the psychology of a stranger" because you would likely be able to interpret a person's personality traits upon the first interaction with a person. In terms of personality traits being cross-situationally consistent, personality psychologists mean that a person's personality traits tend to be expressed across most situations.

To describe personality traits as relatively enduring means that these traits are relatively stable and unchanging across the entire life span (Conley, 1985; McCrae & Costa, 1997, 1999). Personality psychologists call this characteristic of personality traits longitudinal consistency (McCrae & Costa, 1999). Also, personality traits are the dimension of personality that scientists have discovered to be about 50% due to heredity (Bouchard et al., 1990). And personality psychologists have discovered in cross-cultural studies of personality traits that they tend to be universal across cultures, with the exception of the trait of openness to experience and agreeableness. However, there has been some debate about the validity of this claim based on distinctions in cultural values (e.g., individualism vs. collectivism), language, and ethnocentrism within North American psychology.

Most psychologists consider the Big Five personality trait theory (McCrae & Costa, 1997, 1999; Goldberg, 1993; John & Srivastava, 1999) the most scientifically valid model of personality traits even though other models of personality are used within organizations to assess individuals' personality traits. Extroversion, neuroticism, openness, conscientiousness, and agreeableness are the five basic personality traits that, personality psychologists describe, characterize a person's personality. Personality psychologists describe the five personality traits in the following way (John & Srivastava, 1999; McCrae & John, 1992):

Extroversion: The personality trait extroversion includes the following characteristics: (1) activity level (e.g., energetic), (2) dominance, (3) sociability, (4) expressiveness (e.g., showing off), and (5) positive emotionality (spunky). A person high on the personality trait extroversion would be described as a person who is active, assertive, sociable, adventurous, and enthusiastic. A person low on extroversion would be described as more introverted, reserved, and submissive to authority.

Neuroticism: The personality trait neuroticism includes the following characteristics: (1) anxious, (2) self-pitying, and (3) hostile. A person high on neuroticism would be described as having the behavioral tendency to experience strong emotional states such as sadness, worry, anxiety, and fear, as well as overreacting to frustration and being overly irritable, self-conscious, vulnerable, and impulsive. A person low on neuroticism would be described as calm, confident, contented, and relaxed.

Conscientiousness: The personality trait conscientiousness includes the following characteristics: (1) competent (e.g., efficient), (2) orderly, (3) dutiful, (4) self-disciplined, and (5) deliberate. A person high on conscientiousness would be described as a person who is dependable, organized, planful, responsible, and thorough. A person low on conscientiousness would be described as irresponsible and easily distracted.

Agreeableness: The personality trait agreeableness includes the following characteristics: (1) warmth, (2) tendermindedness (e.g., kindheartedness), (3) altruism (e.g., generosity), (4) modesty, and (5) trust. A person high on agreeableness would be described as a person who is appreciative, forgiving, kind, and sympathetic. A person low on agreeableness would be described as shy, suspicious, and egocentric.

Openness: The personality trait openness includes the following characteristics: (1) aesthetically reactive, (2) introspective, and (3) values intellectual matters. A person high on openness would be described as a person who is artistic, curious, imaginative, insightful, and original and has wide interests. A person low on openness would be described as a person who has a pronounced sense of right and wrong.

There are several ways in which psychologists assess a person's personality traits. The most commonly used assessments are the NEO-PI-R, the Big Five Inventory (BFI), and the Myers–Briggs Type Indicator (MBTI). Both the NEO-PI-R and BFI are standardized questionnaires that measure the five major domains of personality traits within the Big Five model of personality traits. One major difference between the two measures is the number of items and cost. The NEO-PI-R includes 240 items, and each assessment has a cost. The BFI includes 44 items and can be used at no cost.

Though most personality psychologists do not consider the MBTI (Briggs & Myers, 1976) a personality trait assessment per se, it is often considered one by those who use it outside of academic research. The MBTI is used to create a typology of personality characteristics based on Jung's (1971) theory of personality types. The personality types are assessed using a four-dimension score based on 95 of 166 items. The four MBTI dimensions are as follows: extraversion–introversion (E–I), sensing–intuition (S–N), thinking–feeling (T–F), and judging–perceiving (J–P). The MBTI is frequently used in applied organizational and assessment settings (Bess & Harvey, 2002). Within these settings, it is used for human resources hiring (i.e., employment selection and hiring), career counseling, and self-development. Even though the MBTI is frequently used in these settings, most academic psychologists regard the MBTI as a measure that has little validity (Reynierse, 2000).

PERSONALITY CHARACTERISTIC ADAPTATIONS AND "THE MOTIVATED AGENT"

A key feature of what it means to be human is to be able to adapt. The typical or characteristic way in which individuals adapt to the various roles, situations, and developmental periods in their lives is essential to how personality operates within the characteristic adaptations dimension of human personality. Characteristic adaptations are contextualized aspects of personhood and include motives, goals, psychological defenses, emotional intelligence, personal concerns, values, and copying style. Unlike personality traits that are relatively fixed across situations and a person's lifetime, personality characteristic adaptations shift or adapt based on social roles, situations, and developmental periods (e.g., childhood, adolescence, adulthood).

Personality psychologists think of a person's motivation as the answer to the question "what energizes and directs a person's behavior?" or more simply "what puts a person's behavior in motion?" McAdams (2015) describes people can engage in different situations, across different roles, and through distinctive developmental periods as a "motivated agent" whose behavior is characterized as a "full sequence of motivated agency. We want something. We set forth the goal of getting what we want. We develop a plan in order to get it. We execute the plan. We achieve the goal" (p. 170). Across the history of the field of psychology, scholars have created many different theories of human motivation (e.g. Banks, McQuater, & Sonne, 1995; McClelland, Atkins, Clark, & Lowell, 1953). However, more contemporary theories of human motivation emphasize the underlying force of agency in directing behavior. Thus, thinking of the person as a motivated agent means that by virtue of being human what the person naturally most desires is to have a plan and believe the plan will work.

There are many different types of motivation and goal pursuits. Each has its own complexities and nuances. However, for sake of simplicity, we will provide a snapshot of the motivation and goal concepts that we believe may be most relevant for understanding the attacker's personality in terms of how it can inform questions about behavior involved in the identification, analysis, and management of cybersecurity events faced by individuals, as well as within organizations.

Extrinsic Motivation versus Intrinsic Motivation

Extrinsic motivation is characterized by a person's behavior being driven by external contingencies. A person then is directed toward rewards (e.g., material gain, prestige, social approval) or avoiding punishments. In contrast, intrinsic motivation is characterized by a person's behavior being driven by the value and reward inherent in the activity itself (e.g., enjoyment).

Power Motivation: Striving for Power

Power motivation is the desire to have an impact on other people, to affect their behavior or emotions, and includes overlapping concepts such as influence, inspiration, nurturance, authority, leadership, control, dominance, coercion, and aggression (Winter, 1992).

Social Motivation: The Striving for Affiliation

Social motivation is the desire to interact with others and gain a sense of affiliation, relatedness, and belongingness. For example, the human need to form and maintain strong, stable interpersonal relationships is belongingness (e.g., Baumeister & Leary, 1995). The psychological mechanisms of enhancement (and protection) of the self, trusting, understanding, and controlling are the foundation of belongingness. These psychological mechanisms facilitate a person being able to effectively function in social groups. Among the core elements of belongingness is the need for affiliation. Heckhausen (1991) described these aspects of social bonding in terms of various forms of social affiliation that include love of family (e.g., children, spouse, parents) and friendship, as well as seeking and maintaining strong relationships with unfamiliar others of the same gender and similar age. Intimacy motivation focuses on the motive to pursue warm, close, and communicative interactions with those with whom an individual has a one-on-one relationship (McAdams, 1985).

Achievement Motivation: The Striving for Achievement

Achievement motivation is the reoccurring desire to perform well and strive for success (McClelland et al., 1953). In other words, most people's behavior is driven by a desire to be effective and competent (McAdams, 2015).

Personality psychologists use several approaches to assess motivation. These include the following:

Thematic apperception test (TAT): Projective tests have been used to measure achievement, power, and social motives (Smith, 1992). A common projective test is the TAT. This test has prompts to which the individual responds to tell a story about a picture. Then the researcher interprets these stories, most often by using motivational themes that have been commonly found within previous motivation research. These themes are most often specified in what researchers call coding schemes or scoring systems. Examples of these coding systems are published in Smith (1992) as follows: the intimacy motivation scoring system (McAdams, 1992), the affiliation motive scoring manual, the affiliative trust–mistrust scoring system (McKay, 1992), the achievement motive scoring manual (McClelland, Atkinson, Clark, & Lowell, 1953), the scoring manual for the motive to avoid success (Horner & Fleming, 1977), and the scoring system for the power motive (Winter, 1992).

Achievement motivation questionnaires are used to measure an individual's motivation. With the exception of the personality research form (Jackson et al., 1996), almost all of these are designed to measure the achievement motivation of students who are at various levels of their development (i.e., children, adolescents, and young adults). A commonly used questionnaire is the Personality Research Form (Jackson et al., 1996; Mayer et al., 2007). This assessment measures achievement motivation along with 21 other types of motives. The *Academic Motivation Scale* (Vallerand et al., 1992) assesses an individual's achievement motivation along, as well as intrinsic and extrinsic motivation. The *Motivated Strategies for Learning Questionnaire* (Pintrich et al., 1993) includes measuring strategies for college students measuring cognitive and motivational components of learning and resource management. The personal strivings assessment (Emmons, 1986) is often used to measure personal goals and their motivational themes including the following: avoidance goals, achievement, affiliation, intimacy, power, personal growth and health, self-presentation, independence, self-defeating, emotionality, generativity, and spirituality (Emmons, 1989).

THE NARRATIVE IDENTITY DIMENSION OF HUMAN PERSONALITY: THE AUTOBIOGRAPHICAL AUTHOR

All humans are autobiographical authors. Across different developmental stages of our lives, people are naturally inclined toward psychological preoccupation to pursue identity questions like, "Who am I?" (Erikson, 1968), "How do I integrate the many roles I play and experiences across time in a meaningful way to have a coherent sense of self?" and "How do I find unity and purpose in life?" (McAdams, 2001). How do people answer these identity questions? They do so through the development of an inner narrative or what psychologists call the narrative identity dimension of human personality. Narrative identity is a person's internalized, integrative, and evolving narrative of self (Singer et al., 2013; Winston, 2011).

Autobiographical memories of selective life experiences serve as building blocks for narrative identity (Singer et al., 2013). Autobiographical memories that are vivid, emotionally intense, and repetitive become self-defining. And as they evolved as self-defining, they connect to a person's other significant memories that share their themes and narrative sequences of the meaning of life experiences. Autobiographical memories reflect a person's most enduring concerns (e.g., achievement, intimacy, spirituality) and/or unresolved conflicts (e.g., sibling rivalry, ambivalence about a parental figure, addictive tendencies) (Singer et al., 2013).

Another way to understand narrative identity is to conceptualize it a life story (e.g., McAdams, 1985, 1988). In their minds, starting during adolescence individuals begin to create a personal life story. The psychological process of autobiographical narration in which a person engages to cultivate a personal life story is largely unconscious and selective. In so doing, the person engages in reinterpretation of the past, perceived present, and anticipated future (McAdams, 2001; Singer, 2004, 2005). In fact, this is one of the features of being human that distinguish us from animals. In contrast to personality traits, narrative identity is the most malleable dimension of human personality. Also, narrative identity is profoundly shaped by the cultural context in which a person's life is embedded.

Psychologists have discovered that the way a person narrates matters. For example, individuals with a narrative identity that finds "redemptive meaning in suffering and adversity and who construct life stories that feature themes of personal agency and exploration, tend to enjoy higher levels of mental health, well-being, and maturity" (McAdams & McLean, 2013, p. 233). Relatedly, Singer et al. (2013) articulated how health narrative identity combines narrative specificity with adaptive meaning-making to achieve insight and well-being. These research findings align with the psychological functions of narrative identity and life story construction to provide a person with a sense of purpose, self-coherence, and identity continuity over time. Narrative identity development can be thought of as a psychosocial journey of being in pursuit of a way to meaningfully articulate a story about the self that integrates multiple roles in the person's life. Selective life experiences across the life course that are embedded with a complex sociocultural context are also integrated into a person's meaning-making process to develop narrative identity (Hammack, 2008; McAdams & Manczak, 2011; McAdams, 2001; McLean et al., 2007; Winston-Proctor, 2018).

Personality psychologists use several approaches to assess the narrative identity dimension of human personality. Among the most popular is interviewing an individual using a semi-structured interview guide (e.g. the life story interview instrument and a guided autobiography instrument (see https://www.sesp.northwestern.edu/foley/). The life story interview involves a 2-hour procedure in which the person provides an interviewer a narrative account of his or her life—past, present, and imagined future—by responding to a series of open-ended questions. The procedure begins by asking the person to divide his or her life into chapters and provide a brief plot outline of each. Next, the interviewer asks for detailed accounts (i.e., autobiographical memories) of eight key scenes in the story, including a high point, a low point, and turning point scene. The interview protocol also includes imagined future chapters and the basic values and beliefs on which the story's plot is developed.

The guided autobiography instrument is similar. However, it does not include interview prompts to get the person to generate life story chapters. Instead, like the life story interview instrument, the guided autobiography asks the person to generate autobiographical memories about each of the following types of key life events or scenes: a peak (high point) life experience, a nadir (low point) life experience, a turning point life experience, a continuity life experience, a childhood life experience, an adolescent life experience, a morality life experience, and a goal-oriented life experience. Using the autobiographical memories generated by the narrative identity instrument, the psychologist then can assess narrative identity by making interpretations about the autobiographical memories reoccurring themes, points of view (i.e. ideological setting), reoccurring characters or archetypes, emotional tone, and narrative complexity.

CONCLUSION

Understanding the multiple dimensions of personality (i.e., personality traits, personality characteristic adaptations, narrative identity) can inform how to identify, analyze, and manage behavior involved in cybersecurity events faced by individuals, as well as within organizations.

Gaining knowledge about the theoretical conceptualization of personality used by personality psychologists allows those interested in behavioral cybersecurity to explore on their own trends in personality development that derive from ongoing scientific research. Also, with this knowledge, individuals and organizations can explore how these dimensions of personality are revealed from clues manifested from observation of cybersecurity cases and scenarios.

REFERENCES

Banks, W., McQuater, G., & Sonne, J. (1995). A Deconstructive Look at the Myth of Race and Motivation. *The Journal of Negro Education*, 64(3), 307–325.

Baumeister, R. F., & Leary, M. R. 1995. The need to belong: Desire for interpersonal attachments as a fundamental human motivation. *Psychological Bulletin*, 117(3), 497–529.

Bess, T. L., & Harvey, R. J. 2002. Bimodal score distributions and the Myers-Briggs type indicator: Fact or artifact? *Journal of Personality Assessment*, 78(1), 176–186.

Bouchard, T. J., Jr., Lykken, D. T., McGue, M., Segal, N. L., & Tellegen, A. 1990. Sources of human psychological differences: The Minnesota study of twins reared apart. *Science*, 250, 223–228.

Briggs, K., & Myers, I. 1976. *The Myers-Briggs Type Indicator*. Consulting Psychologists Press, Palo Alto, CA.

Conley, J. J. 1985. Longitudinal stability of personality traits: A multitrait–multimethod–multioccasion analysis. *Journal of Personality and Social Psychology*, 49, 1266–1282.

Emmons, R. A. 1986. Personal strivings: An approach to personality and subjective wellbeing. *Journal of Personality and Social Psychology*, 51, 1058–1068.

Emmons, R. A. 1989. The personal striving approach to personality. In L. A. Pervin (Ed.), *Goal Concepts in Personality and Social Psychology*. Lawrence Erlbaum Associates, Hillsdale, NJ, pp. 87–126.

Erikson, E. H. 1968. *Youth and Crisis*. Norton, New York.

Goldberg, L. R. 1993. The structure of phenotypic personality traits. *American Psychologist*, 48, 26–34.

Hammack, P. L. 2008. Narrative and the cultural psychology of identity. *Personality and Social Psychology Review*, 12, 222–247.

Heckhausen, H. 1991. Social bonding: Affiliation motivation and intimacy motivation. In H. Heckhausen (Ed.), *Motivation and Action*. Springer, Berlin, Heidelberg, pp. 301–316.

Horner, M. S., & Fleming, J. 1977. *Revised Scoring Manual for an Empirically Derived Soring System for the Motive to Avoid Success*. Unpublished manuscript, Harvard University, Cambridge, MA.

Jackson, D. N., Paunonen, S. V., Fraboni, M., & Goffin, R. D. 1996. A five-factor versus six-factor model of personality structure. *Personality and Individual Differences*, 20, 33–45.

John, O. P., & Srivastava, S. 1999. The big five trait taxonomy: History, measurement, and theoretical perspectives. In L. Pervin & O. P. John (Eds.), *Handbook of Personality: Theory and Research* (2nd ed.). Guilford Press, New York, pp. 102–138.

Jung, C. G. 1971. *Psychological Types*. Princeton University Press, Princeton, NJ. (Original work published 1923.)

Mayer, J. D., Faber, M.A., & Xu, X. 2007. Seventy-five years of motivation measures (1930–2005): A descriptive analysis. *Motivation Emotion*, 31, 83–103.

McAdams, D. P. 1985. *Power, Intimacy, and the Life Story: Personological Inquiries into Identity*. Guilford Press, New York.

McAdams, D. P. 1988. *Power, Intimacy and the Life Story*. Guilford Press, New York.

McAdams, D. P. 1992. The intimacy motivation scoring system: Handbook of thematic content analysis. In Motivation and *Personality*: Handbook of *Thematic Content Analysis* (pp. 229–253). Cambridge University Press, Cambridge.

McAdams, D. P. 2001. The psychology of life stories. *Review of General Psychology*, 5(2), 100–122.

McAdams, D. P., and Pals, J. L. (2006). A new Big Five: Fundamental principles for an integrative science of personality. *American Psychologist*, 61(3), 204–2127.

McAdams, D. P. 2015. *The Art and Science of Personality Development*. Guilford Press, New York.

McAdams, D. P., & Manczak, E. 2011. What is a "level" of personality? *Psychological Inquiry*, 22(1), 40–44.

McAdams, D. P., & McLean, K. C. 2013. Narrative identity. *Current Directions in Psychological Science*, 22(3), 233–238.

McClelland, D. C., Atkinson, J. W., Clark, R. A., & Lowell, E. L. 1953. *Century Psychology Series. The Achievement Motive*. Appleton-Century-Crofts, East Norwalk, CT.

McCrae, R. R., & Costa, P. T., Jr. 1997. Personality trait structure as a human universal. *American Psychologist*, 52, 509–516.

McCrae, R. R., & Costa, P. T., Jr. 1999. A five-factor theory of personality. In L. Pervin & O. John (Eds.), *Handbook of Personality: Theory and Research*. Guilford Press, New York, pp. 139–153.

McCrae, R. R., & John, O. P. 1992. An introduction to the five-factor model and its applications. *Journal of Personality*, 60(2), 175–215.

McKay, J.R. 1992. A coring system for affiliative trust-mistrust. In C. P. Smith (Ed.), *Motivation and Personality: Handbook of Thematic Content Analysis*. Cambridge University Press, New York, pp. 266–277.

McLean, K. C., Pasupathi, M., & Pals, J. L. 2007. Selves creating stories creating selves: A process model of self-development. *Personality and Social Psychology Review*, 11, 262–278.

Patterson, W., & Winston-Proctor, C. E. 2019. An international extension of Sweeney's Data Privacy Research. Advances in Human Factors in Cybersecurity, T. Ahram & W. Karwowski (Eds.),

Proceedings of the AHFE 2019 International Conference on Human Factors in Cybersecurity, July 24–28, Washington DC, pp. 28–37.

Pintrich, P. R., Smith, D. A., Garcia, T., & McKeachie, W. 1993. Reliability and predictive validity of the Motivated Strategies for Learning Questionnaire (MSLQ). *Educational and Psychological Measurement*, 53, 801–813.

Reynierse, J. H. 2000. The combination of preferences and the formation of MBTI types. *Journal of Psychological Type*, 52, 18–31.

Singer, J. A. 2004. Narrative identity and meaning making across the adult lifespan: An introduction. *Journal of Personality*, 72(3), 437–459.

Singer, J. A. 2005. *Personality and Psychotherapy: Treating the Whole Person*. Guilford Press, New York.

Singer, J. A., Blagov, P., Berry, M., & Oost, K. M. 2013. Self-defining memories, scripts, and the life story: Narrative identity in personality and psychotherapy. *Journal of Personality*, 81, 569–582.

Smith, C. P. 1992. *Motivation and Personality: Handbook of Thematic Content Analysis*. Cambridge University Press, Cambridge.

Vallerand, R. J., Pelletier, L. G., Blais, M. R., & Brière, N. M. 1992. The academic motivation scale: A measure of intrinsic, extrinsic, and motivation in education. *Educational and Psychological Measurement*, 52, 1003–1017.

Winston, C.E. 2011. Biography and life story research. In S. Lapan, M. Quartaroli & F. Riemer (Eds.) *Qualitative Research: An Introduction to Designs and Methods*. Jossey-Bass, Hoboken, NJ, pp. 106–136.

Winston-Proctor, C. E. 2018. Toward a model for teaching and learning qualitative inquiry within a core content undergraduate psychology course: Personality psychology as a natural opportunity. *Qualitative Psychology*, 5(2), 243–262.

Winter, D. 1992. Power motivation revisited. In C. Smith (Ed.), *Motivation and Personality: Handbook of Thematic Content Analysis*. Cambridge University, Cambridge, pp. 301–310.

4

HACKER CASE STUDIES
Personality Analysis and Ethical Hacking

In studying various persons who have spoken or written about their exploits as hackers or cybersecurity criminals, it seems that we can gain some knowledge about their technical achievements by analyzing their personalities.

One person who has been well known in the hacker community for many years is Kevin Mitnick. Mitnick enjoyed great success in compromising telephone networks (phone phreaking), until he went to federal prison in the 1980s. Subsequently, he has become an influential writer and speaker about cybersecurity. One of his best known books is *The Art of Intrusion: The Real Stories Behind the Exploits of Hackers, Intruders and Deceivers*, written with William Simon (Mitnick & Simon, 2005).

This book details the exploits of numerous hackers, alleged to be true stories, and in many of the cases describes enough about the interests and motivations of the subjects to give us some insight into their personalities.

We select just a few examples from *The Art of Intrusion* to identify some of the personalities and what drove them to the exploits described in the book.

COMRADE

The hacker known as Comrade began his exploits as a teenager living in Miami. About some of his early works, Comrade said, "we were breaking into government sites for fun." Comrade developed a friendship with another hacker with an Internet name of neoh, another young man who was only a year older than Comrade, but who lived 3000 miles away. About his interests, neoh said, "I don't know why I kept doing it. Compulsive nature? Money hungry? Thirst for power? I can name a number of possibilities." Also, neoh, in corresponding with author Mitnick, wrote: "You inspired me ...

I read every possible thing about what you did. I wanted to be a celebrity just like you."

Another person, named Khalid Ibrahim, who claimed to be from Pakistan, began to recruit Comrade and neoh. Khalid's interest was in working with other hackers who might be willing to hack into specific targets—first in China and then in the United States. Khalid indicated that he would pay cash for the successful penetration into the targets he indicated.

Comrade's interest, as he indicated, was that he knew that Khalid

> was paying people but I never wanted to give up my information in order to receive money. I figured that what I was doing was just looking around, but if I started receiving money, it would make me a real criminal.

ADRIAN LAMO

Adrian Lamo, as a teenager, lived in New England and developed his hacking skills at an early age.

Mr. Lamo, according to his parents, was involved in hacking because of a number of specific well-known hackers who were his inspiration. His strategy in hacking was to understand the thought processes of the person who designed the subject of his attacks, a specific program or network. In one case, he discovered a customer who asked for assistance with stolen credit card numbers, and the technicians who are supposed to assist did not bother responding. Then Adrian called the victim at home and asked if he had ever gotten a response. When the man said no, Adrian forwarded the correct answer and all the relevant documentation regarding the problem. As Lamo said, "I got a sense of satisfaction out of that because I want to believe in a universe where something so improbable as having your database stolen by somebody… can be explained a year later by an intruder who has compromised the company you first trusted."

Adrian's description of his philosophy can be summarized as: "I believe there are commonalities to any complex system, be it a computer of the universe … Hacking has always been for me less about technology and more about religion."

GABRIEL

Gabriel lives in a small town in Canada, and his native language is French. Although he sees himself as a white hat hacker, he occasionally commits a malicious act when he finds a site "where security is so shoddy someone needed to be taught a lesson." As a young man, he found details about the IP addresses of a small bank in the US south that nevertheless had extensive national and international ties. He discovered that one of the bank's servers ran software that allows a user to remotely access a workstation. Eventually he found ways to remotely access terminal service, so he could essentially own the potential system. He also found the password for the bank's firewall, and so hacking into that one machine gave access to other computer systems on the same network. As a consequence, Gabriel had access to great deal of internal information, but he did not have any interest in stealing funds.

In addition to relating individual hacker personalities, a number of authors have attempted to describe generic personality traits of hackers.

HACKER PERSONALITY DESCRIPTIONS

Lee Munson is a security researcher for Comparitech and a contributor to the Sophos' Naked Security blog. Munson has written (Munson 2016),

> It's hard to pin down just a few personality traits that define a hacker. A typical hacker profile is a male, age 14–40, with above-average intelligence, obsessively inquisitive with regards to technology, non-conformist, introverted, and with broad intellectual interests. A hacker is driven to learn everything he can about any subject that interests him.
>
> In fact, most hackers that excel with technology also have proficiency in no technological hobbies or interests. Hackers tend to devour information, hoarding it away for some future time. Credit card and bank fraud present opportunities to use cracking to increase personal wealth.

Eric Stephen Raymond is the cofounder of the Open Source Initiative, an organization that builds bridges between the hacker community and business. Raymond has written (Raymond 2015):

> Although high general intelligence is common among hackers, it is not the sine qua non one might expect. Another trait is probably even more important: the ability to mentally absorb, retain, and reference large amounts of 'meaningless' detail, trusting to later experience to give it context and meaning. A person of merely average analytical intelligence who has this trait can become an effective hacker. In terms of Myers-Briggs and equivalent psychometric systems,

hackerdom appears to concentrate the relatively rare INTJ and INTP types; that is, introverted, intuitive, and thinker types.

Rick Nauert has over 25 years' experience in clinical, administrative, and academic healthcare. He is currently an associate professor for Rocky Mountain University of Health Professionals doctoral program in health promotion and wellness. And Nauert likens the personality traits of hackers with the symptoms of autism (Nauert 2016).

> Online hacking costs the private and corporate sectors more than $575 billion annually. While security agencies seek out "ethical" hackers to help combat such attacks, little is known about the *personality traits* that lead people to pursue and excel at hacking.
>
> New research shows that a characteristic called systemizing provides insight into what makes and motivates a hacker. Intriguingly, the personality traits are similar to many autistic behaviors and characteristics ... Systemizing is the preference to apply systematic reasoning and abstract thought to things or experiences. The preference for systemizing is frequently associated with autism or Asperger's, a milder form of autism.

ETHICAL HACKING

A hacker is a person with the technical skill and knowledge to break into computer systems, to access files and other information, to modify information that may be in the computer system, to utilize skills involving network technology to move from one system to another, and to implant software that may have deleterious effects on the host system.

An ethical hacker is a person with the technical skill and knowledge to carry out the same functions as aforementioned, but to resist doing so for ethical reasons.

Given the fact that there is essentially no difference in the technical skill set of a hacker or an ethical hacker, one might wonder what difference in fact can there be. It is also the case that another terminology has become widespread: The hackers are often called "black hats" and the ethical hackers "white hats." However, to further confuse the issue, there are competitions wherein the participants are assigned to the white hat team or the black hat team, but in midcompetition, they may change hats and change roles.

This terminology is somewhat unusual. In other areas of human activity where we consider behavior as being either legal or illegal, ethical or unethical, it stretches the imagination to consider: bank robbery or ethical bank robbery; murder or ethical murder. If Robin Hood were real, he would probably like to have been considered an ethical thief.

A distinguished computer scientist named Ymir Vigfusson, originally from Iceland and more recently a professor at Emory University in Atlanta, offers courses in Ethical Hacking and has described his philosophy extremely well in a recent Ted Talk "Why I Teach People How To Hack" (https://youtu.be/KwJyKmCbOws) (Vigfusson 2015).

Prof. Vigfusson uses the term "moral compass" to describe what guides him as a professor in teaching ethical hacking to his students and also how he operates in his own practice.

Thus, it seems that the challenge in the cybersecurity profession is to find a way of identifying how an individual who can develop the requisite technical skills can rely on his or her own moral compass. Developing measures to try to predict these behaviors is a clear challenge for those persons who are not only knowledgeable about cybersecurity but also about psychology and the behavioral sciences.

PROGRAMS TO ENCOURAGE THE DEVELOPMENT OF ETHICAL HACKERS

Very recently, greater attention has been drawn to initiatives that attempt to encourage the development of ethical hackers.

Donna Lu wrote in *The Atlantic* (Lu, 2015):

> The cybersecurity expert Chris Rock is an Australian information-security researcher who has demonstrated how to manipulate online death-certification systems in order to declare a living person legally dead.
>
> Rock began researching these hacks last year, after a Melbourne hospital mistakenly issued 200 death certificates instead of discharge notices for living patients. He also uncovered similar vulnerabilities in online birth registration systems. The ability to create both birth and death certificates meant that hackers could fabricate new legal identities.

Subsequently, on August 2, 2017, Kevin Roose in the *New York Times* wrote (Roose 2017):

> If there is a single lesson Americans have learned from the events of the past year, it might be this: hackers are dangerous people. They interfere in our elections, bring giant corporations to their knees,

and steal passwords and credit card numbers by the truckload. They ignore boundaries. They delight in creating chaos.

But what if that's the wrong narrative? What if were ignoring a different group of hackers who aren't lawless renegades, who are in fact patriotic, public-spirited Americans who want to use their technical skills to protect our country from cyber-attacks, but are being held back by outdated rules and overly protective institutions?

In other words: What if the problem we face is not too many bad hackers, but too few good ones?

And most recently, on November 24, 2017, Anna Wiener wrote in the *New Yorker* (Wiener, 2017):

> "Whenever I teach a security class, it happens that there is something going on in the news cycle that ties into it," Doug Tygar, a computer-science professor at the University of California, Berkeley, told me recently. Pedagogically speaking, this has been an especially fruitful year. So far in 2017, the Identity Theft Resource Center, an American nonprofit, has tallied more than eleven hundred data breaches, the highest number since 2005. The organization's running list of victims includes health-care providers, fast-food franchises, multinational banks, public high schools and private colleges, a family-run chocolatier, an e-cigarette distributor, and the U.S. Air Force. In all, at least a hundred and seventy-one million records have been compromised. Nearly eighty-five per cent of those can be traced to a single catastrophic breach at the credit-reporting agency Equifax.

PROBLEMS

1. Read *The Art of Intrusion*. Identify any of the characters portrayed as female, or any of the characters described (or that you would estimate) as being over 50 years in age.
2. Critique the description Lee Munson has provided in this chapter of the personality traits of a hacker.
3. Research the Myers–Briggs personality types indicator system (see the Chapter 3 on personality tests, also https://upload. wikimedia.org/wikipedia/commons/1/1f/MyersBriggsTypes. png). Identify categories unlikely to be attributed to a hacker.
4. Discover if there are any professions that seem to have an overabundance of persons with Asperger's syndrome.
5. Watch the YouTube and Ted Talk by Ymir Vigfusson. What caused him joy to discover his hack and then respond to his "moral compass"?
6. Find the origin of the terms "white hat" and "black hat."

REFERENCES

Lu, D. 2015. "When Ethical Hacking Can't Compete," *The Atlantic*, December 8, 2015.

Mitnick, K. & Simon, W. 2005. *The Art of Intrusion: The Real Stories Behind the Exploits of Hackers, Intruders and Deceivers*. Wiley Publishing, Indianapolis, IN.

Munson, L. 2016. Security-FAQs. http://www.security-faqs.com/what-makes-a-hacker-hack-and-a-cracker-crack.html.

Nauert, R. 2016. *PsychCentral.com*. https://psychcentral.com/news/2016/06/02/some-personality-traits-of-hackers-resemble-autism/104138.html.

Raymond, E. S. 2015. *Catb.org*. http://www.catb.org/jargon/html/appendixb.html.

Roose, K. 2017. "A Solution to Hackers? More Hackers," *New York Times*, August 2, 2017.

Vigfusson, Y. 2015. "Why I Teach People How to Hack," *Ted Talk*, March 24, 2015.

Wiener, A. 2017. "At Berkeley, A New Generation of 'Ethical Hackers' Learns to Wage Cyberwar," *New Yorker*, November 24, 2017.

5

PROFILING

There is a technique in law enforcement called "profiling," which has been used over many years in order to determine if a given criminal behavior leads to a mechanism for defining a category of suspects.

Although this approach has been successful in many instances, it has also led to widespread abuses. First, let us consider one of the major types of abuse. The expression "driving while Black" has evolved from the all-too-common practice of police officers stopping an automobile driven by an African-American who might be driving in a predominantly white neighborhood. As eminent a person as former President Barack Obama has reported that this has happened to him on occasion (Soffen, 2016).

Thus in the realm of cybersecurity, it is important to recognize that criteria that might be used to develop profiling approaches need to be sensitive in attributing certain types of behavior based on race, ethnicity, or other identifiable traits of an individual, rather than attributing types of behavior based on actions rather than individual traits.

PROFILING IN THE CYBERSECURITY CONTEXT

There has been a growing body of instances of cyberattacks where there is a great need to try to isolate one or several potential perpetrators of the attack.

Throughout this book, you will see a number of case studies of both actual and fictional cyberattacks, and the profiling techniques described here may be applicable in determining potential suspects. To introduce the subject, however, we will use one well-known series of incidents that we might describe as the "Sony Pictures Hack."

SONY PICTURES HACK

In October 2014, word leaked out that Sony Pictures had under development a film titled *The Interview*. What was known was that the storyline for this film was that the US government wanted to employ journalists to travel to North Korea to interview the president of that country—and under the cover of the interview, to assassinate the president.

As word of this plot leaked out, the North Koreans—understandably furious—threatened reprisals for Sony and for the United States should this film be released.

A *hacker* group which identified itself by the name "Guardians of Peace" (GOP) *leaked* a release of confidential data from Sony. The data included personal information about Sony employees and their families, emails between employees, information about executive salaries at the company, and copies of then-unreleased Sony films. This group demanded that Sony pull its film *The Interview* and threatened *terrorist attacks* at cinemas screening the film. After major US cinema chains opted not to screen the film in response to these threats, Sony elected to cancel the film's formal premiere and mainstream release, opting to skip directly to a digital release followed by a limited theatrical release the next day.

US intelligence officials alleged that the attack was sponsored by North Korea.

Sony was made aware of the hack on Monday, November 24, 2014, as malware previously installed rendered many Sony employees' computers inoperable by the software, with the warning by GOP, along with some of the confidential data taken during the hack. Several Sony-related accounts were also taken over, and several executives had received via email on the previous Friday, coming from a group called "God'sApstls" [*sic*], demanded "monetary compensation." Soon the GOP began leaking yet-unreleased films and started to release portions of the confidential data to attract the attention of social media sites.

Other emails released in the hack showed Scott Rudin, a film and theatrical producer, discussing the actress Angelina Jolie very negatively.

On December 16, for the first time since the hack, the GOP mentioned the then-upcoming film *The Interview* by name and threatened to take terrorist actions against the film's New York City premiere at Sunshine Cinema on December 18, as well as on its national release date of December 25. Sony pulled the theatrical release the following day.

> We will clearly show it to you at the very time and places The Interview be shown, including the premiere, how bitter fate those who seek fun in terror should be doomed to. Soon all the world will see what an awful movie Sony Pictures Entertainment has made. The world will be full of fear. Remember the *11th of September 2001*. We recommend you to keep yourself distant from the places at that time. (If your house is nearby, you'd better leave.)

The stars of *The Interview* responded by saying they did not know if it was definitely caused by the film but later canceled all related media appearances.

Undoubtedly all of the publicity involved with these threats, and the embarrassing disclosures from the Sony hack, led to a great deal of public interest in the film and the resulting controversy. As a result, many people, who would have been not willing to take a risk by going to the theater showing this film, decided to purchase and view it in the safety of the streamed video.

Thus, *The Interview* became somewhat of an underground hit, and it is not disputed that many more people saw the film because of the controversy.

The question thus became who might have been the perpetrator of the Sony hack, and the resulting actions related to the controversy regarding the release of the film.

PROFILING MATRICES

With the information given regarding the Sony Hack, we can use the information from the series of events to develop a model to try to narrow the consideration of potential perpetrators of this hack as well as their motivations for carrying this out.

Rather than having an approach that will conclusively lead to a definitive answer, the profiling matrix approach will provide the researcher with a means to narrow the potential suspects and the corresponding motivations to a point where many suspects may be effectively eliminated.

First, we will build a list of potential suspects. The beginning step here is to gather as much information as possible regarding candidates to be included or rejected as suspects. From the case study as presented earlier, most persons would immediately include the North Korean government as a suspect. Along those lines, however, it may be that rather than the North Korean government itself which may or may not have the requisite technical ability—actors on their behalf might be considered. As we have said, a suspected hacker group that calls itself the GOP had claimed that they had executed the Sony hack. However, GOP had not been previously identified, so it is possible that they were only a pseudonym for some other hacker group. The group Anonymous might also be suspect, since they had claimed, with some reason, that they had perpetrated numerous other hacks in the past.

It is also known through any analysis of world politics that China is a country that has the strongest working relationship with North Korea and is also known to have very substantial technical capabilities.

But there had also been suspicion related to a technically skilled Sony employee, who is referred to as Lena in numerous articles. Lena had been fired by Sony not long before the hack in question, and it was widely known that she had the capability to perform the hack and may have wished revenge.

It is also reasonable to consider in any competitive environment that competitors of Sony might have also had a motivation to do damage to Sony's reputation.

At the point where there was considerable discussion as to whether or not the film *The Interview* might be pulled from being released, or released in a fashion that might diminish its profitability, persons that stood to benefit from the success of the film—for example, the producer and director, or the lead actors, might have a motive in terms of either decreasing or increasing the value of the film.

A final consideration might be that Sony Pictures themselves might have realized that creating considerable controversy over the release of the film, which otherwise might have gone mostly unnoticed, could result in greater profitability for Sony themselves.

When offering a course at Howard University, and using the Sony Pictures Hack as an example, the complete list of 16 potential suspects became:

North Korea

Guardians of Peace

Iran

Sony Employees

WikiLeaks

Russia

China

Microsoft

Industrial Competitor

Movie Industry

Google

Anonymous

Lena

MGM

Seth Rogen, James Franco

Sony Pictures

We now have a potential list of suspects, our next task is to try to identify the reasons for the motivations for this series of events. It is conceivable that one could identify a new motivation for which none of the previous suspects (identified by roles) might be identified. Nevertheless, money and politics are obvious.

Anytime we identify groups for whom hacking is a primary activity, we should consider if the hack is carried out to demonstrate technical proficiency—or even to try to "advertise" for future business. Revenge is often another motive, certainly in this case given the existence of the disgruntled former employee. And industrial competitiveness is often another potential motivation.

Perhaps one more candidate should be added to this list: Given the results that Sony was much more profitable with this film after all of the controversy, it might be considered that stirring the controversy could have been perceived as a potential for making the series of events beneficial to Sony itself.

Thus, we identified 12 possible motivations as follows:

Politics

Keep the peace

Warfare

Reputation

Start conspiracy

Become famous

Personal vendetta

Money

Disclose information

Adventure

Abuse

Competition

Our next step is now to create the columns of the profiling matrix. Thus, we list all of the potential motivations that we have identified and now we can create a matrix whose 16 rows are labeled by the suspects and the 12 columns labeled by the potential motivations.

Thus, we have defined the columns of our profiling matrix.

The next step in the analysis is to examine each cell in this newly formed matrix and then, based on all of the documentary evidence at hand for the case, estimate the probability that this particular cell, defined by a pair (perpetrator, motivation), might be the most likely guilty party and rationale. The results in our classroom exercise are shown in Figure 5.1.

In Chapter 9, the method of establishing profiling matrices is extended in a transformation to a game theory model that allows for a solution to be developed from the profiling data.

THE "ABCD" ANALYSIS

Another methodology for analyzing levels of threat in a cybersecurity environment is the so-called "ABCD" approach. This approach, which had its origins in criminal justice theory, attempts

Motivations ->	Politics	Keep the peace	Warfare	Reputation	Start conspiracy	Become famous	Personal vendetta	Money	Disclose Information	adventure	Abuse	Competition
Suspects												
North Korea	43.21	2.50	31.07	18.21	10.00	1.07	10.36	4.29	7.14	0.36	0.36	0.36
Guardians of Peace	12.86	10.36	4.29	41.07	16.43	23.21	9.64	3.21	11.43	3.93	1.07	0.71
Iran	21.26	1.19	12.45	14.12	13.41	2.26	3.76	5.12	6.07	1.07	1.07	0.36
SONY Employees	1.07	11.79	0.89	6.79	3.57	6.43	34.64	20.18	21.25	11.79	2.32	1.07
WikiLeaks	27.90	4.33	4.29	7.65	15.08	1.43	0.00	1.07	41.60	3.02	0.00	1.15
Russia	23.54	0.79	36.70	2.74	30.99	1.94	0.71	5.00	6.62	2.30	0.71	1.87
China	35.78	3.06	22.8	1.02	28.28	3.06	2.14	7.86	3.21	0.00	0.71	0.71
Microsoft	4.12	4.95	1.55	10.55	7.26	0.36	0.00	28.36	5.00	0.00	0.00	15.00
Industrial Competitor	3.57	3.57	0.71	13.57	12.86	1.79	2.86	48.21	2.86	2.14	2.86	15.36
Movie Industry	0.71	7.86	0.00	3.57	6.43	1.43	1.43	15.00	3.57	0.00	0.00	22.86
Google	3.39	9.29	0.89	3.39	5.18	0.00	0.00	16.61	9.64	5.00	0.89	7.14
Anonymous	3.89	1.07	1.07	7.53	17.45	1.38	6.07	4.59	31.54	25.39	2.45	3.65
Lena	0.36	2.14	0.00	17.12	3.81	4.29	30.33	23.14	10.24	0.00	0.00	0.00
MGM	1.43	4.29	0.00	7.14	10.00	0.71	0.00	11.43	0.00	0.00	0.00	14.29
Seth Rogen, James Franco	1.79	13.21	1.79	1.43	2.86	19.29	5.00	20.71	1.79	0.00	0.00	4.29
SONY Pictures	3.93	20.17	2.74	2.14	17.95	5.95	0.00	29.02	2.38	0.00	0.00	8.57

Figure 5.1 Profiling matrix for the Sony Hack.

to define an attack in terms of its sophistication. As a simplification, potential cyberattacks or the hackers developing or utilizing these attacks are divided into four categories, with the least sophisticated attacks categorized as "D-level"; going all the way to the most sophisticated attacks were attackers who fall into the "A-level" category. Our examples will include both car theft and cyberattacks (Table 5.1).

TABLE 5.1
Classification of Malware: The "ABCD" Model for Bad Guys

Level of Criminal	Description	Car Theft Example	Cybersecurity Example
D	Smash-and-grab: no skill, knowledge, resources, sees opportunity, and acts immediately	A potential thief walks along a street where there are a number of parked cars and comes to a car where the thief sees a briefcase sitting on the passenger seat, observes the presence of a large rock beside the sidewalk, grabs it, and immediately throws it at the window, shattering it and therefore allowing the person to take the briefcase and run	Script kiddies: the D-level cybercriminal is a person who is pointed to a website that might contain embedded software that can be downloaded and run in order to create some form of attack. The effort requires no skill or knowledge but is simply acted upon once the malicious software is found. Sometimes these perpetrators have been called "script kiddies"
C	Some limited skills and resources, but little planning in execution	The potential thief's objective is actually to steal the car. Since there are many cars on the street, the C thief looks for a target. If the thief noticed a club steering wheel lock, he or she would probably move on to the next car, since the thief's limited skills would cause the realization that destroying the club might occupy enough time to expose the thief to capture	Low orbit ion cannon: the C-level cybercriminal might be one who is able to launch a form of a DDoS (Distributed Denial of Service) attack. For example, by downloading the program "Low Orbit Ion Cannon," one can simply type in the URL and thus launch various types of the DDoS attacks

(Continued)

TABLE 5.1 (*Continued*)
Classification of Malware: The "ABCD" Model for Bad Guys

Level of Criminal	Description	Car Theft Example	Cybersecurity Example
B	Very knowledgeable, some resources, ability to plan	There is in fact a well-developed "university" system for the training of the B level criminal: It is called the prison system. Where except in prison can one obtain—without the cost of tuition—the highest level of training in committing criminal acts, with therefore the greatest knowledge but yet a small amount of resources? This thief might know how to "hotwire" a car to be able to start it up, and take off	Internet worm: the B-level cybercriminal might be someone with a great deal of programming skill, sufficient to create a worm, virus, or ransomware and to launch it to attack some well-chosen target site or sites
A	Well-organized team, very sophisticated knowledge, lots of resources, only interested in large targets, can plan extensively	The leader or part of an organization that is in the business of receiving the stolen car that might be delivered by the B-level accomplice, but then have the knowledge, resources, organization, and planning to be able to strip down these vehicles and repackage them for future sales	Stuxnet: The A-level cybercriminal might in fact be a government or a vast organization sponsored by government or governments. Examples of A-level attacks might be Stuxnet or variants. Stuxnet has been reliably identified as a joint project of both the US and Israeli governments

PROBLEMS

1. Find an example beyond "Driving While Black" with the generic assumption of identifying with an ethnic or racial stereotype.
2. Identify some values and some dangers of profiling.
3. Can you find a tragic outcome of profiling?
4. Assess the reviews of movie *The Interview* related to the Sony Hack.
5. Has "Lena" ever been identified?
6. Identify your own candidates as suspects and motivations for the Sony profiling case.
7. Construct an ABCD description for the classification of potential bank robbers.
8. Use the ABCD method to classify the following:
   A, B, C, or D:
   A. pickpocket
   B. fabrication of student grades
   C. the Morris worm
   D. script kiddies

REFERENCE

Soffen, K. 2016. "The Big Question about Why Police Pull Over So Many Black Drivers," *Washington Post*, July 8, 2016. https://www.washingtonpost.com/news/wonk/wp/2016/07/08/the-big-question-about-why-police-pull-over-so-many-black-drivers/?utm_term=.7346a524986f.

6

ACCESS CONTROL

ACCESS CONTROL

Our first step in protecting a computing environment or cyberenvironment is to establish methodologies for determining how access may be gained to our environment. We usually divide this concern into two components that we call authentication and authorization.

Our concern in providing authentication is basically to answer the question "Who are you?" In other words, this means the establishment of a mechanism for determining whether a party wishing to gain access is allowed to enter the system. In this case, the party in question might be either a human or a machine, the authentication process is initiated by that external party, and our system must respond appropriately.

The second aspect of access control is called authorization. In other words, once an external party has been authenticated, questions may arise as to whether that party has the authority to perform certain tasks in this new environment. In other words, the authorization question might be "Are you allowed to do that?" And so, our system must have a methodology for enforcing limits on actions.

AUTHENTICATION

The process of authentication begins with a request from the external party, followed by a challenge from our system, which usually can be divided into one of three approaches:

Something you know
Something you have
Something you are

SOMETHING YOU KNOW: PASSWORDS

The something you know is usually thought of as a password. Something you have may be some physical device such as a key, smart card, or some other token. And something you are is usually described as a biometric, in other words, your fingerprints, the image of your face, the scan of your retina, your method of walking or gait, or your DNA, among others.

Thus, there is a long history of users creating good or bad passwords, and this has been one of the biggest problems in the world of cybersecurity. The concept is that the user chooses a good password—in other words, one that is hard for an outsider to guess—and that will foil an outsider from making a successful guess.

Every password system has allowable types of symbols or characters to be typed. Examples are digits { 0, 1, 2, ..., 9 }, letters of the alphabet (lowercase) {a, b, c, ..., z}, special (typable) symbols {#, $, %, ^, &, ?, !}, or combinations of these. The set of these that is being used will be designated as c, for the character set.

The second part is how many of these symbols may or must be typed in for a legitimate password. For the moment, consider that this must be a fixed number, n.

It is important to know, therefore, how many possible passwords there can be. Since there are c choices for each entry, and there must be n of them, the total number of potential passwords is c^n.

Example: For many ATM pins, four digits are required, and there are 10 potential digits. So, c = 10, n = 4, and the total number of possible pins is $c^n = 10^4 = 10,000$.

Example: In many older password systems, seven characters needed to be entered, each one a lowercase letter. Thus, $c^n = 26^7 = 8,031,810,176 = 8.03 \times 10^9$ or just over 8 billion possible passwords.

Let us call this value the "password set," $p = c^n$. This value also indicates the challenge to someone trying to obtain someone else's password.

Since the system itself will usually instruct a potential user as to the password rules, the hacker trying to determine a user's password will know c and n and thus can calculate p. So the most obvious hacker's approach is usually called "brute force"—try all the potential passwords.

GOOD PASSWORD CHOICE

How to create good passwords is a long-standing, but nevertheless still perplexing, problem for computer users everywhere. Undoubtedly this problem will remain until such time as the community of computer users determines that passwords are not an appropriate test for authentication—for example, to be replaced by biometrics (what you are) or physical devices (what you have).

On the other hand, if one assumes that we must commit a password to memory, then we must be confident that our memory is sufficient to contain such a password—or, in general in these modern times, contain the multiple passwords that we need for not only our computing account, but also access to many other websites such as our bank account, our accounts with bills that we may have to pay electronically, or online vendors where we may purchase various merchandise.

With the use of multiple passwords, of course our human memory requirements multiply as well.

The suggestion that we tend to use is to choose a relatively short, but memorable event, so that it is indelibly burned into our conscious, but also one that is not well known to persons who wish to research our histories. Let us say that many years ago, I had a memorable encounter of a person, place, or thing—let us say it was "Penobscot," which is actually a river in the state of Maine.

Furthermore, in this case, I had told no one of having spent some time along that particular river—so no one who knew me would make such an association. So now, this is a password I am unlikely to ever forget; to use Penobscot but isolate from a dictionary attack, I might insert some other character into the password so that simply by testing for the password with a dictionary of river names, we might deflect such an attack by using as the password, for example, penob7scot.

PASSWORD METERS

Another approach to the creation and use of secure passwords is through one of a number of so-called "password meters" that are available at various sites on the Internet.

In particular, we have examined five of these candidates for password meters. In each case, the meter is available through a particular website, the user is encouraged to enter a test password, and a report is generated for the user as to the password meter's judgment of the strength of the password. We designate them as (A) https://passwordmeter.com; (B) https://lastpass.com/howsecure.php; (C) https://my1login.com/resources/password-strength-test/; (D) https://thycotic.com/resources/password-strength-checker/; and (E) https://howsecureismypassword.net/.

Unfortunately, what we have too often discovered is that the strength of password judged by any one of the test password meters may vary completely from the judgement by one or several of the other candidates for password meters.

One way of testing the validity for consistency of a proposed password meter is to submit a number of prospective passwords to each meter candidate and examine the consistency of the results:

Test Password	Feature
11111111111111111111	A rather simple but very long string; a nuisance to type
Penob7scot	A word found in some types of dictionaries, for example, a dictionary of US place names, with the insertion of a number to defeat a dictionary attack
x3p9q!m	A seemingly random string, but difficult to remember
Brittttany	A person's name, with a certain letter duplicated (T); probably easy to remember
Onomatopoeia	A long word, but in most dictionaries
aBc123xYz	Follows a pattern, but the mixture of letters, numbers, and capitals could prove strong and in this case probably easy to remember

The results of the meters (A) through (E) on the test passwords were as follows:

A	B	C	D	E
6 Very weak 0%	2 Weak	6 Very weak 0.04 seconds	3 2 weeks	1 79 years
2 Strong 63%	2 Moderately strong	2 Strong 7 months	1 3 years	2 8 months
3 Good 54%	4 Weak	1 Strong 8 months	6 1 minute	6 22 seconds
5 Very weak 8%	2 Moderately strong	5 Weak 16.05 minutes	5 4 hours	5 59 minutes
4 Very weak 13%	1 Very strong	3 Medium 13 hours	2 4 months	3 4 weeks
1 Strong 76%	4 Weak	4 Weak 40.01 minutes	3 2 weeks	4 4 days

TOKENS: WHAT YOU HAVE

Devices that you physically possess are becoming less and less common in contemporary computing environments. The reason is simply that if you have some form of key, token, smartcard, or other physical device, and if it is simply lost or stolen and only that physical device is necessary for entry or authorization in some environment, then essentially the door is left wide open.

Consequently, more and more the physical device is used as part of what is called "two-factor authorization." Now, more often, the physical device is combined with a second factor, normally a password.

A more clever device is the so-called RSASecurID security token (CDW, 2018), which is synchronized in a multiuser system with the time that the token is activated. This device has a six-digit display, which changes every 60 seconds. The user with the token must enter the displayed six digits concurrently with the entry of a password. The central system uses an algorithm that can determine to the minute what the token's six-digit readout should display, and the user must enter the six digits to match what the system stores.

More and more in recent times, websites that require that you use a password mechanism to have an account at the website (banks, credit cards, web retailers, or e-tailers) use a two-factor authentication system. Typically, you create a password that the e-tailer then stores. Periodically, you might have to change the password—or more likely, address the issue if you happen to forget the password. In this case, if you indicate you wish to change or restore your password, the system will ask for a second account—typically a cell phone number—where you will find an email or phone message with a secondary password in order to update your password or other security information.

BIOMETRICS: WHAT YOU ARE

The field of biometrics has been in existence much longer than the computer era. Perhaps one of the most common biometric measurements—what you are—is the fingerprint. The study of classification of humans by fingerprint dates back to the nineteenth century. Other more contemporary biometric measurements include facial recognition, hand recognition, retinal patterns, and DNA.

PROBLEMS

1. Which of the following pairs of password schemes provides the greater security in terms of the largest number of legal passwords in the scheme?
   a. U: (c, n) = (26, 8) or V: (c, n) = (20, 9)
   b. W: (c, n) = (52, 8) or X: (c, n) = (72, 7)
   c. Y: (c, n) = (62, 13) or Z: (c, n) = (20, 18)
2. Suppose the password set size is $10^9 = 1$ billion. Find a minimal value for c when n = 7. Find a minimal value for n when c = 52.
3. Suppose we have the following password authentication schemes. Indicate the solutions for each case in the matrix below, assuming a brute force attack.

Case	Scheme	Total Number of Attempts Necessary to Find a Password	Expected Number of Attempts Necessary to Find a Password
A	Exactly six alphanumeric characters, case sensitive		
B	Any word in the Oxford English Dictionary		
C	Fifteen characters, which all must be vowels (i.e., a, e, i, o, u, y)		
D	A four-digit PIN (numeric characters only)		
E	Six characters, which must have at least one in each category: non–case-sensitive letters, numbers, and one of the following special symbols {! # $ % ^ & * ~ – /}. ("{ }" represents a set notation, not two of the special characters)		
F	Exactly ten alphabetic characters, non-case sensitive		

4. Consider the RSA token. With a six-digit display, changing every minute, how long will it take until the display repeats?

REFERENCE

CDW Inc. 2018. RSA SecureID. https://www.cdw.com/search/?key=rsa%20securid%20sid800&searchscope=all&sr=1.

7

THE FIRST STEP
Authorization

Now we assume that an external user has satisfied the requirements for authentication and has entered a system. Although for many of us, we may have a personal computer with access only for one person, more and more the norm in a cyberenvironment is that there may be multiple parties present in the same environment with distinct sets of resources (files or applications).

Thus, the problem of authorization must be addressed. Suppose a user requests access to a specific file. What is the mechanism to ensure that that particular user has the authorization to either read, write, or delete that file, no matter where in this environment this resource resides. Over time, there have been many approaches to this problem.

SECURITY LEVELS

Complex modern environments tend to have multiple ways of representing the level of access going beyond the familiar UNIX model. Many of us will be familiar with the specific levels of access in a government or military system. In perhaps its simplest form, these levels can be described as Top Secret, Secret, Confidential, and Unclassified.

In such a system, these levels are applied to each subject and object. Then the manner of accepting or rejecting a specific request will follow this hierarchy:

Top Secret > Secret > Confidential > Unclassified

Systems implementing such a hierarchy are normally called multilevel security models (MLSs). An early method of implementing such a model was the so-called Bell–LaPadula model (BLP) (Bell & LaPadula, 1973). In the simplest form of this model, the rules are as follows: (1) classifications apply to objects and (2) clearances apply to subjects.

The BLP security model is designed to express the essential requirements for an MLS. Primarily, BLP deals with confidentiality, in order to prevent unauthorized reading. Recall that O is an object, with a classification, and that S is a subject, with a clearance.

The security level is denoted L(O) for objects and L(S) for subjects. The BLP rules consist of the

> *Simple Security Condition*: S can read O if and only if $L(O) \leq L(S)$
>
> **-Property (Star Property)*: S can write O if and only if $L(S) \leq L(O)$

As a shorthand term, this is often referred to as "no read-up, no write-down."

PARTIAL AND TOTAL ORDER

It is normal in many authorization systems for the user to have not only a security level, as discussed earlier, but also a secondary level of authorization involving a subset of the users at a given security level. For example, a number of users with Secret clearance may be assigned to work together on a single project, but all other users with Secret clearance not involved in this project have no need for the information that is contained in the development of the project, i.e., the "need-to-know" principle. Thus, in this case, a user working on the project will by necessity not only have Secret clearance but also need to be a member of the project group. Therefore, the complete security status for anyone in this environment will require both the level of clearance and the listing of the user's groups.

Thus, when a request is made by a subject to access a given object, the system must check both respective security levels. The overall security level is described as a "total order." In other words, it is always possible to determine the access: A Confidential user may never read a Secret object, or a Secret user may write to a Top Secret object.

Consider this example of a project where the teams are divided into groups called {Red, Blue, Green, Yellow, Brown}, user A belongs to groups Red, Green, Brown, and object B belongs to groups Blue, Green, Yellow. Then neither a request for A to read B

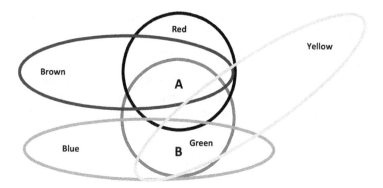

Figure 7.1 Venn diagram for {Red, Blue, Green, Yellow, Brown}.

nor a request for A to write to B will be honored, because the subset for A is not contained in the subset for B, nor vice versa (Figure 7.1).

COVERT CHANNEL

Despite the protections involved in the access control system, there can always be the existence of a mechanism to "get around" these protections. These are typically what are called "covert channels." Covert channels are often defined as communication paths that are not intended as such by system designers.

Here is one example that could exist in a multilevel security system. Suppose that a user, George, has Top Secret clearance, and another user, Phyllis, has Confidential clearance. Furthermore, suppose that the space for all files is shared. In this case, George creates a file called ABC.txt, which will be assigned Top Secret level since it was created by George with that clearance. Of course, Phyllis cannot read that file, having only Confidential clearance. However, in order for George to communicate a covert message bit by bit to Phyllis, Phyllis checks George's directory every minute. If in the first minute, Phyllis sees the existence of ABC.txt, she records a "1" bit. If, 1 minute later, George deletes the file and Phyllis checks again—by not seeing the name of the file in the directory—she will record a "0." Then, George may create the file again, so in the next minute, Phyllis will see the existence of the file and record a "1." Thus, over a period of, say, 60 minutes, George can "leak" essentially a message of 60 bits.

Consider one further example of a covert channel. Assume that we have a 100-MB Top Secret file, stored in its unencrypted fashion at the Top Secret level. However, the encrypted version may be stored at the Unclassified level for the purposes of transmitting it from one system to another. This presents no security risk because we would assume that the encrypted version is useless to someone discovering this version without the key. However, the key, stored at the Top Secret level, is perhaps just a few hundred bits. Thus, a method such as that described earlier, leaking one bit at a time from the Top Secret level to the Unclassified level, could make the complete key available at the Unclassified level in just a few minutes and pass on the ability to decrypt the 100-MB file.

INFERENCE CONTROL

A major application for many is use and access in a major database, such as a corporate database, or perhaps a university database. The purpose of a database is to allow for multiple users to read or modify information from the database according to access rules such as those described earlier. Any user with the ability to submit a database query may be able to inadvertently (or perhaps intentionally) gain information that the user may be prohibited from accessing directly. Consider the following example from a university database:

Question: What is average salary of female psychology professors at XYZ University?

Answer: $95,000

Question: How many female psychology professors are there at XYZ University?

Answer: 1

As a result of these two queries, specific information has leaked from responses to general questions!

Inference Control and Research

Medical records are extremely private but critically valuable for epidemiological research. The question is: How can we make aggregates of this medical information available for epidemiological research and yet protect an individual's privacy? Or, equivalently, how can we allow access to such data without leaking specific information?

A NAÏVE ANSWER TO INFERENCE CONTROL

We could remove names from medical records. Unfortunately, it still may be easy to get specific info from such "anonymous" data, as has been extremely well documented by LaTanya Sweeney (2000), who proved that approximately 87% of the United States' population can be identified as individuals with only three pieces of widely available data, {gender, birthdate including year, and five-digit ZIP code of residence}.

Therefore, removing names is not enough.

RANDOMIZATION

You can add a small amount of random noise to data. However, this works with fields like weight, age (such as 45.2 years), or height but could not be used with ZIP code, or gender (represented by a 0 or 1)— in other words, values on a continuous scale.

FIREWALLS

A firewall must determine what to let into (or out of) the internal network. In other words, the firewall is designed to provide access control for a network. You might think of a firewall as a gate-keeper. To meet with an executive, first contact the gatekeeper. The gatekeeper decides if the meeting is reasonable; also, the gatekeeper filters out many requests.

PROBLEMS

1. A system contains the following:
   Subjects (with classifications and compartments):

Subject	Classification	Compartments
John	Top Secret	{Accounting, Operations, Development}
Mary	Secret	{Accounting, Operations, Production, Marketing}
David	Confidential	{Production, Development}
Ann	Unclassified	{Accounting, Operations, Marketing, Development}

Objects (with clearances and compartments):

Object	Clearance	Compartments
Payroll	Top Secret	{Accounting, Operations, Production}
Inventory	Secret	{Operations, Production, Marketing}
Shipping	Confidential	{Accounting, Operations, Production, Development}
Media	Unclassified	{Accounting, Operations, Production, Marketing}

With a Bell–LaPadula mandatory access control system, indicate whether each of these requests will be accepted or rejected:

Subject	Command	Object	Accept or Reject
John	Read	Inventory	
Ann	Write	Payroll	
Mary	Read	Media	
Ann	Read	Shipping	
David	Write	Shipping	
John	Write	Media	
David	Write	Shipping	
Mary	Read	Inventory	

2. We have a universe U of objects that are the letters of the alphabet, $U = \{a, b, c, ..., z\}$. In this universe, we have objects (words) that define subsets whose elements are the letters that make up the word. For example, the subset for the word elephant is Elephant = $\{a, e, h, l, n, p, t\}$. Construct the Venn diagram that displays the sets corresponding to the words Onomatopoeia, Penicillin, Comatose, Lugubrious, and Syzygy.

3. The US Department of Defense, in understanding that covert channels can never be completely eliminated, established a guideline to allow covert channels whose capacity is no more than 1 bit/second. Suppose a malicious leaker adheres to these guidelines. What is the shortest period of time that she can take to leak a 10-kB file?

4. Inference control: You can find a good deal of US city population data at https://factfinder.census.gov/faces/tableservices/jsf/pages/productview.xhtml?src=bkmk. Formulate a query that will give you all cities with a population < 100,000 at the last census but > 100,000 at present.

REFERENCES

Bell, D. E., & LaPadula, L. J. 1973. Secure Computer Systems: Mathematical Foundations. MITRE Technical Report 2547, Volume I, March 1.

Sweeney, L. 2000. *Simple Demographics Often Identify People Uniquely.* Carnegie Mellon University, Data Privacy Working Paper 3, Pittsburgh, PA.

8

ORIGINS OF CRYPTOGRAPHY

All of the preceding sections have led us to what is perhaps the most important component of creating secure environments. This is the study and application of methods to transform information by a secret function so that it will be extremely difficult for an attacker to be able to interpret the message in question.

The approach that is central to this and any course on cybersecurity is known as cryptology, which is in effect a combined term stemming from the study of cryptography (from the Greek κρυπτο [krypto or *hidden*] and γραφια [grafia or *writing*]). The second component of cryptology is the field of cryptanalysis or the study of ways of breaking cryptographic methods. Both fields have an extremely long history far predating the computer era. In fact, one of the earliest cryptographic methods is said to have been invented by Julius Caesar over 2000 years ago.

CAESAR SHIFT

In the Caesar shift, as well as in any cryptographic method or algorithm, we have a text that is required to be hidden, called the plaintext. We also have a number of different possible ways of transforming that text, each one of those ways determined by the choice of the transformation that we call the key. Once the transformation determined by the key is applied to the plaintext, the result is referred to as the encrypted text or the ciphertext.

In the example of the Caesar shift, suppose that we wish to transmit the message: "ROME WAS NOT BUILT IN A DAY." And, for this transmission today, we choose as the key the number 3, by which we mean that each letter in the message will be advanced by three positions in the alphabet—in the case of letters near the end of the alphabet, if we advance beyond the letter Z, we continue back to the beginning or A. Thus, our plaintext message becomes:

```
ROME WAS NOT BUILT IN A DAY
SPNF XBT OPU CVJMU JO B EBZ
TQOG YCU PQV DWKNV KP C FCA
URPH ZDV QRW EXLOW LQ D GDB
```

So, the ciphertext to be transmitted to the receiver is "**URPH ZDV QRW EXLOW LQ D GDB**".

In order for any cryptographic method to be useful, there must be an inverse to the key in order to allow the receiver to apply that inverse key and thus retrieve the original message. In the case of the Caesar shift, the inverse key is the inverse of the value of the key in arithmetic modulo 26—or, in other words, the number that when added to the key (3) adds to 26, in other words $3 + 23 \equiv 0 \pmod{26}$.

Thus, the method of decrypting the encrypted text is to use the same process, in this case by advancing each letter 21 times. Of course, the equivalent is to back each letter in the encrypted text up three times. Again, we use the "cyclic" alphabet: after Z we go to A.

In order to use this or any other cryptographic method, we require a method for formulating plaintext; a number of possible transformations, each defined by a key, and a system or alphabet for the ciphertext. Finally, we require that the choice of the key or transformation be kept secret, but somehow it must also be shared between the sender and the receiver of the encrypted message so that the receiver can apply the inverse key in order to retrieve the original plaintext.

In the case of the Caesar shift, one failing is that there are a maximum of 26 possible keys, one for each letter of the alphabet, so for the attacker trying to break the Caesar shift, it is only necessary to try all 26 potential keys, and one of them will produce the desired result.

SUBSTITUTION AND TRANSPOSITION

Most cryptographic methods over time have relied on two philosophically different approaches to the transformation of information. These are called substitution and transposition. The Caesar example cited earlier uses the principle of substitution; in other words, each character in the plaintext is transformed according to a function of the character symbol itself. In the case of our aforementioned example with the key "5," the 26 letters of the alphabet are transformed as follows: A → F, B → G, C → H, ..., Z → E.

The other approach, transposition, which we will see later, performs the transformation on the position of the symbol in the plaintext list. A very simple example might be described as (1 2 3 4) → (3 4 2 1), by which we mean that the first of a four-letter group moves to the third position, the second to the fourth, and so on. Thus, in this case, the message "EACH GOOD DEED PAYS" becomes "HCEA DOGO DEDE SYPA." The inverse key is therefore (1 2 3 4) → (4 3 1 2).

THE KEYWORD MIXED ALPHABET CRYPTOSYSTEM

The keyword mixed alphabet uses for its keys the set of all words, with duplicate letters removed, in the English language. Indeed, the requirement for words to be English words is imposed only because the distribution of the keyword is much simpler if it is a word rather than an arbitrary character string.

> The messenger, having ridden from Lexington to Valley Forge in half a day, was exhausted, out of breath, and indeed near death as he approached General Washington to tell him the secret key for the cipher system: "XRUTGDKWQFP", he panted, then expired. Did he say "XRUTGDKWQFP" or "XRUTGDKWQFT"? puzzled General Washington.

The method itself uses the keyword to define a mapping or permutation of the message space alphabet. To define the key, the alphabet is written in its normal order, and under it is written a permuted alphabet, with the letters of the keyword followed by the remaining letters of the alphabet. For the moment, consider only that we are using just the 26 uppercase characters.

Suppose the keyword is: FACETIOUSLY

Then the regular alphabet is (with the spaces inserted just for readability):

ABCDE FGHIJ KLMNO PQRST UVWXY Z

and the permuted alphabet is:

FACET IOUSL YBDGH JKMNP QRVWX Z

Written for ease of reading, we have:

```
ABCDE FGHIJ KLMNO PQRST UVWXY Z
FACET IOUSL YBDGH JKMNP QRVWX Z
```

The encryption maps each letter of the message text to a letter of cipher text according to the permutation defined earlier. Thus, "MAY

THE FORCE BE WITH YOU" becomes "DFX PUT IHMCT AT VSPU XHQ," or, more likely, "DFXPU TIHMC TATVS PUXHQ." It is common practice to group the text in equal size blocks. On the one hand, the spaces allow for easier human reading, but more importantly, the normal position of the blanks or spaces tells us the word lengths.

THE VIGENÈRE CRYPTOSYSTEM

The Vigenère cipher was a widely used cryptosystem dating back to the sixteenth century, using a keyword combined with a Caesar shift. If the keyword is "FACETIOUSLY," an 11-letter word, as before, the encryption will use 11 different Caesar shifts periodically. (Each letter determines a Caesar shift, or modular addition.) Suppose that $0 \leftrightarrow A$, $1 \leftrightarrow B$, ..., $25 \leftrightarrow Z$, as usual. Then, the first letter to be encoded uses the shift corresponding to F, the second to A, the third to C, and so on until the cycle repeats:

Choose a key word, perhaps:

Plain text:

IT'S A LONG WAY TO TIPPERARY ...

Key:

FAC E TIOU SLY FA CETIOUSLY

Cipher text:

OUV F FXCB PMX ZP WNIYTMTDX

ONE-TIME PAD ENCRYPTION

The "one-time pad" plays an important role in the history of encryption. It was first described in 1882 by Frank Miller.

The principle of the one-time pad is an elaboration on the Vigenère cryptosystem as described earlier. In the Vigenère system, the key word is chosen (FACETIOUSLY in our example), and then encryption consists of a different Caesar shift based on the corresponding element in the pad corresponding to the plaintext letter in the same position. However, the breaking or cryptanalysis of the Vigenère method is based on the realization that after a certain number of plaintext letters are encrypted, the period related to the key may be determined. In the case of the example with keyword FACETIOUSLY, the same Caesar shift will take place with characters in positions 1, 12, 23, 34, and so on: in other words, every 11th letter.

The concept with the one-time pad is that the length of the key word can be stretched to an enormous length. Indeed, for the true one-time pad, the keyword should be of infinite length. However, the creation of an infinite-length string that does not repeat is practically impossible. But it is possible using randomization techniques to generate long sequences of letters that may only repeat after thousands or indeed millions of terms in a sequence.

Long after one-time pads began to be used in military applications, it was proved by Claude Shannon of Bell Laboratories in 1949 (Shannon, 1949) that the one-time pad could be proved in its original description as the most secure encryption possible, despite the fact that the true one-time pad of infinite length is essentially impossible to construct.

For example, one possible one-time pad might be:

```
S   I   S   A   V   C   U   I   I   D   F   Z   D
M   M   M   Y   G   L   X   S   K   G   K   S   I
J   C   A   H   O   O   F   N   C   W   F   L   V
L   Z   R   M   G   Y   V   G   I   E   D   Q   A
O   X   J   K   I   Y   W   G   E   N   D   I   D
V   W...
```

So, let us use this quote from Albert Einstein to encrypt:

The difference between stupidity and genius is that genius has its limits.

THE DIFFERENCE BETWEEN STUPIDITY AND GENIUS IS THAT GENIUS HAS ITS LIMITS

```
THEDI   FFERE   NCEBE   TWEEN   STUPI   DITYA   NDGEN
IUSIS   THATG   ENIUS   HASIT   SLIMI   TS


SISAV   CUIID   FZDMM   MYGLX   SKGKS   IJCAH   OOFNC
WFLVL   ZRMGY   VGIED   QAOXJ   KIYWG   EN
DIDVW   (etc.)
```

Using the correspondence A ↔ 1, B ↔ 2, C ↔ 3, ..., Z ↔ 26, add the value for each plaintext character to the value of the corresponding element of the pad (in mod 26, or subtracting 26 if the sum is >26). The encryption of these characters is the character corresponding to the sum.

```
T → 20, S → 19; 20 + 19 = 39; remainder mod 26 =
13, 13 → M
H →  8, I →  9;  8 +  9 = 17; remainder mod 26 =
17, 17 → Q
E →  5, S → 19;  5 + 19 = 24; remainder mod 26 =
24, 24 → X
```

And so on.

THE PLAYFAIR SQUARE

Here is a more sophisticated method from the nineteenth century known as the Playfair Square. To create a Playfair Square, we will build a 5×5 square using the regular English alphabet (Patterson 1987).

Whoops! The alphabet has 26 letters and yet 5×5 = 25.

Our task is to reduce the alphabet to 25 letters, which we accomplish by merging two letters together, I and J. So, in the algorithm, in the plaintext, every J is changed to an I.

Now pick a word to form the key. If the keyword has a duplicate, for example, suppose the keyword is CONSPIRATORIALLY, then by deleting, we will use CONSPIRATLY as the key.

Now create the Playfair Square by writing the letters of the key into the 5×5 square in row order. Then fill in the rest of the (25-letter) alphabet, also in row order. The Playfair Square encryption method is based not on substituting single letters (therefore 26 different substitutions) when taking letters in pairs (or digrams). There are 26×25 = 650 digrams; therefore, the total number of possible permutations of those 650 symbols is a number approximately 8.1×10^{1547}, that is, greater than a 1 followed by 1547 zeros. In this case, the Playfair Square is:

```
C  O  N  S  P
I  R  A  T  L
Y  B  D  E  F
G  H  K  M  Q
U  V  W  X  Z
```

We will refer to this square as row 1, "**CONSP**"; Column 1, "**CIYGU**"; and so on.

The 5×5 square is essentially the key, although all that has to be shared with the receiver is the keyword, CONSPIRATORIALLY in this case (Figure 8.1).

Now to encrypt a message, the square is based on the symbols being not letters but pairs of letters. We first break the plaintext message into digrams. For example, the plaintext message COMMITTEE becomes:

CO MM IT TE EZ (if the plaintext has an odd number of letters, just attach any letter at the end).

Encrypt rule 1: If the digram is in the same row or column, shift each character to the right (for a row) or down (for a column); thus CO → ON, IT → RL.

Encrypt rule 2: If the digram defines a rectangle, take the other corners of the rectangle, in the order of the row of the first message character; thus, for example, EZ → FX.

But, for our keyword CO MM IT TE EZ, how do we deal with MM? The answer is: we don't. We preprocess the message by inserting a letter (Q is best) between every pair. Thus, actually COMMITTEE becomes COMQMITQTEQE or CO MQ MI TQ TE QE. Thus, we have ensured that in the text to be encrypted, we will have no double letters and thus either rule 1 or rule 2 will apply. The encryption of the new plaintext becomes

Encryption:	Rule:	Location:
CO → ON	Rule 1	Row 4
MQ → QG	Rule 2	Row 1
MI → GT	Rule 2	Rows 4,1
TQ → LM	Rule 2	Rows 2,4
TE → EM	Rule 1	Column 4
QE → MF	Rule 2	Rows 4,3

Now you can send the message. The receiver has the same square and has two encrypt rules, rule 1 going left if the pair is in the same row, going up if they are in the same column, and the rectangle rule 2 is the same as for encryption.

Finally, as the receiver has the decrypted but slightly altered message, the receiver pulls out all the Qs inside pairs and interprets "I" as "J" where necessary.

When we change all Js into Is, this will almost never confuse issues in the decryption. And, with respect to the inserted Qs between double letter pairs, can this ever become a problem? Since in English, a Q is almost always followed by a U, this will almost never create a problem—except possibly for VACUUM.

ROTOR MACHINES

Early in the twentieth century, machine production of ciphers became possible. To mechanize the production of ciphertext, various devices were invented to speed the process. All important families of such devices were rotor machines, developed shortly after the time of World War I, to implement Vigenère-type ciphers with very long periods.

A rotor machine has a keyboard and a series of rotors. A rotor is a rotating wheel with 26 positions—not unlike the workings of an odometer in an automobile. Each position of the rotor wheel completes an electric contact and, depending upon the position, determines a different Caesar shift. When a key on the keyboard is depressed, a letter is generated dependent on the position of the rotors.

WORLD WAR II AND THE ENIGMA MACHINE

A US patent for a rotor machine was issued in 1923 to Arthur Scherbius (IT History Society, 2018). It essentially formed the basis for the German Enigma machine. A variation on the rotor machine design was developed by the German Armed Forces leading up to World War II and called the Enigma machine (Figure 8.1). The now well-documented British success at breaking the code at Bletchley Park was a major factor in the war effort. In 2014, the movie *The Imitation Game* depicted reasonably well the efforts led by Alan Turing in breaking the Enigma code (Sony Pictures Releasing, 2014). Turing, considered by many the father of computer science, died tragically in 1954 at age 41.

PROBLEMS

1. Find two strings of the greatest length that make sense in English and are related by a Caesar shift. (Example: t4(CAP) = GET. Thus, t1(BZO) = CAP and t5(BZO) = GET. Ignore blanks.) If the "CAP/GET" example is length 3, the record in our classes is length 7.
2. What is unique about the keyword FACETIOUSLY?
3. Decrypt the following cipher using Caesar's shift:

   RVPXR WPCXV JTNQR VJWXO ONAQN LJWCA NODBN

4. Using the (1 2 3 4 5) → (4 3 5 1 2), encrypt "FOURSCORE AND SEVEN YEARS AGO."

Figure 8.1   The Enigma machine orrectshown at the Swiss Transport Museum. (Creative Commons: Enigma_Verkehrshaus_Luzerne.jpg licensed with CC-by-sa-3.0.)

5. Decrypt this message that uses a transposition cipher:

   RTEH IANI ANPS SIIN NMIA NLIY PTEH NLIA

6. Using the Vigenère cipher method with a ten-letter keyword of your choice, encrypt the message:

   I have a dream that one day on the red hills of Georgia the sons of former slaves and the sons of former slave owners will be able to sit down together at the table of brotherhood.

7. Create a Playfair Square with keyword IMPOSSIBLE and encrypt the message JUSTICE EVENTUALLY WINS OUT.

8. Billary and Hill use a Playfair cipher to prevent their communications from being intercepted by an intruder, Trumpy. They have agreed on a common key, from the word "ALGEBRAICALLY."

   a. Construct their jointly used Playfair Square.

A	L	G	E	B
R	I	C	Y	D
F	H	K	M	N
O	P	Q	S	T
U	V	W	X	Z

  b.   Billary receives the following Playfair-encrypted message
       from Hill:

       RP GX HI PG   BS SY RT RV TO SO
       ML AT SP QW MG SG VH KB

Find Billary's decryption.

REFERENCES

IT History Society. 2018. Dr. Arthur Scherbius Bio/Description. https://
    www.ithistory.org/honor-roll/dr-arthur-scherbius.
Patterson, W. 1987. *Mathematical Cryptology*. Rowman and Littlefield,
    Totowa, NJ, 318 p.
Shannon, C. 1949. Communication theory of secrecy systems. *Bell
    System Technical Journal*, 28(4), 656–715.
Sony Pictures Releasing. 2014. *The Imitation Game* (film).

9

GAME THEORY

Please see edition equations online at https://www.routledge.com/Behavioral-Cybersecurity-Fundamental-Principles-and-Applications-of-Personality/Patterson-Winston-Proctor/p/book/9780367509798.

An important technique in the area of cybersecurity can be in the application of what we know as game theory (Morris, 1994; Von Neumann & Morgenstern, 1944).

For the purposes of this course, we will develop just an introductory exposition to game theory, notably what is now referred to as "two-person, zero-sum games." The term *zero-sum* refers to the constraint that whatever one party gains in such a game, his or her opponent loses. Perhaps the simplest example of such a two-person zero-sum game involves the toss of a coin. The sides of the coin are labeled heads or tails, and one player tosses while the other guesses the outcome of the toss. Let's call the players Mary and Norman and the results of the toss either H or T. Mary tosses the coin, and while it is in the air, Norman guesses the outcome. Clearly, there are four potential results in this game. In two-person game theory, we describe such a game by a matrix, with the rows of the matrix representing the choices for the first player and the columns the choices for the second. In this game, there are only two choices for each player, so we establish a 2 × 2 matrix.

PAYOFF

The "payoff" or outcomes are represented by values in the body of the 2 × 2 matrix. If we assume that if the first player guesses the result of the second player's toss, he or she wins one unit of currency, say one dollar, then the entire game can be described as in this matrix.

		Norman (Column Player)	
	Choice	H	T
Mary (row player)	H	+1	−1
	T	−1	+1

By convention, we always label the row player's positive outcome with a positive number; the column player interprets that value as a loss. So, in the aforementioned example, if Mary tosses heads and Norman guesses tails, then Mary wins a dollar. We enter a −1 in the T row for Norman—which Mary interprets as a +1 for her outcome.

More generally speaking, in such a game, each player knows in advance this payoff matrix, but game theory provides the analytical approach to enable each player to determine his or her best strategy.

In this example, the coin tosser does not have a strategy, but we could alter the game very slightly by saying that rather than conducting a toss, Mary would simply secretly choose one of her outcomes (H or T), and Norman would guess at the result, and then they would compare their choices.

In this model, each would develop a strategy to determine their choice, but a quick analysis would show that there is no winning strategy, as each player as a 50–50 chance of winning.

Suppose, however, that we create a game where the winning strategies might be different for each player. For example, with two players, each will call out either the number one or two. One player is designated as "odd"—let us say he is Oliver—the other player is "even," and she will be known as Evelyn. If the result of the sum of each player's call is odd, Oliver wins that amount; if the result of the sum is even, Evelyn wins that amount. Using the model described earlier for the game matrix, we have:

		Evelyn (Even)	
	Choice	One	Two
Oliver (odd)	One	−2	3
	Two	3	−4

Rather than thinking of any game as a one-time event, consider what players would do if the game may be repeated. Each tries to develop a strategy whereby they can improve their chances. One "strategy" might be simply a random choice of their play, in which case in the aforementioned example, Oliver would choose "one" 50% of the time and "two" the other 50%. Evelyn would reason similarly.

In such a case, Oliver would lose two dollars half the time and win three dollars the other half if he chose "one"; in other words, $0.5(-2) + 0.5(3) = +0.5$.

On the other hand, if Oliver chose "two" half the time, his result would be $0.5(3) + 0.5(-4) = -0.5$. So, from Oliver's perspective, he should just choose "one" all the time. But Evelyn, being no fool, would discern the strategy, choose "two" all the time, and thus win three dollars at each play. Therefore, in order for a game such as this to be competitive, each player must determine a strategy or a probability of each potential choice.

Suppose that Oliver chooses "one" 60% of the time (3/5) and "two" 40% of the time (2/5); then this strategy (assuming Evelyn plays randomly) will net him $-2(3/5) + 3(2/5) = 0$ when he calls "one," and when he calls "two," he wins $3(3/5) - 4(2/5) = 1/5$. So, he breaks even over time when he calls "one" but wins 0.5 when he calls "two."

Suppose we let p represent the percentage of times that Oliver calls "one." We would like to choose p so that Oliver wins the same amount no matter what Evelyn calls. Now Oliver's winnings when Evelyn calls "one" are $-2p + 3(1 - p)$, and when Evelyn calls "two," $3p - 4(1 - p)$. His strategy is equal when $-2p + 3(1 - p) = 3p - 4(1 - p)$.

Solving for p gives 7/12, so Oliver should call "one" $7/12 = 58.3\%$ of the time, and "two" $5/12 = 41.7\%$ of the time. Then, on average, Oliver wins $-2(7/12) + 3(5/12) = 1/12$, or 8.3 cents every time, no matter what Evelyn does. This is called an equalizing strategy.

However, Oliver will only earn this if Evelyn does not play properly. If Evelyn uses the same procedure, she can guarantee that her average loss is $3(7/12) - 4(5/12) = 1/12$. Because each is using their best strategy, the value 1/12 is called the value of the game (usually denoted V), and the procedure each player uses to gain this return is called the optimal strategy or the "minimax strategy."

What we have just described is a mixed strategy. Where the choices among the pure strategies are made at random, the result is called a mixed strategy. In the game we have just described, the pure strategies "one" and "two" and the mixed strategy lead to the optimal with probabilities of 7/12 and 5/12.

There is one subtle assumption here. If a player uses a mixed strategy, he or she is only interested in the average return, not caring about the maximum possible wins or losses. This is a drastic assumption. Here we assume that the player is indifferent between receiving

$1 million for sure and receiving $2 million with probability one-half and zero with probability one-half. We justify this assumption arising from what is called utility theory (see Chapter 16). The basic premise of utility theory is that one should evaluate a payoff by its utility or usefulness and not its numerical monetary value. A player's utility of money is not likely to be linear in the amount.

A two-person zero-sum game is said to be a finite game if both strategy sets are finite sets. Von Neumann has developed the fundamental theorem of game theory, called the Minimax Theorem, which states in particular: For every finite two-person zero-sum game, (1) there is a number V, called the value of the game; (2) there is a mixed strategy for player I such that I's average gain is at least V no matter what player II plays; and (3) there is a mixed strategy for player II such that II's average loss is at most V no matter what player I plays.

If V is zero, we say the game is fair; if V is positive, we say the game favors player I; and if V is negative, we say the game favors player II.

MATRIX GAMES

To this point, we have only considered two-person games where there are only two pure strategies. Clearly, this is a severe restriction: in general, a player will have many pure strategy options.

More generally, a finite two-person zero-sum game can be described in strategic form as (X, Y, A). In this case, X equals the choice among m pure strategies for Player I, $X = \{x_1, x_2, ..., x_m\}$ and Y represents the n pure strategies for Player II, $Y = \{y_1, y_2, ..., y_n\}$. With this terminology, we can form the payoff or game matrix with rows and columns corresponding to the choices of each player. **[See "eresources: Equations and Problems" on the Website mentioned at the beginning of the chapter.]**

In this form, Player I chooses a row i, Player II chooses a column j, and II pays I the entry in the chosen row and column, a_{ij}. The entries of the matrix are the winnings of the row chooser and losses of the column chooser.

MIXED STRATEGY

For player I to have a mixed strategy means that there is an m-tuple p with m distinct probabilities for the choices, $p = \{p_1, p_2, ..., p_m\}$, with the additional restriction that the sum of the p_i is 1.

We also denote the strategy for player II through the n-tuple of probabilities for player II's choices. Again, the sum of the q_j probabilities must be 1.

The representation by the matrix A is a static description of the game. It merely gives the result if the row player chooses strategy x_i and the column player chooses y_j, as the result or payoff is a_{ij}.

However, using matrix algebra, and considering the range of strategies available to each, the probabilities of choice by the row player can be described by a row vector of strategies $p = \begin{bmatrix} p_1 & p_2 & \cdots & p_n \end{bmatrix}$, and the column player's strategies by a column vector $\begin{bmatrix} q_1 & q_2 & \cdots & q_n \end{bmatrix}^T$ (where T represents the transpose, simply an easier way of writing the column player's strategies).

The result of the pure strategy can be described as a special case of the mixed strategy by describing in the pure strategy case; the probability vector for the pure strategy is of the form $\begin{bmatrix} 0 & 0 & \cdots & 1 & \cdots & 0 \end{bmatrix}$, with the 1 in the ith position if the pure strategy is to choose the ith option, and similarly for the column player.

Given either a pure or a mixed strategy, our objective is to "solve" the game by finding one or more optimal strategies for each player. **[See "eresources: Equations and Problems" on the Website mentioned at the beginning of the chapter.]**

The row player's strategies are $p = \begin{bmatrix} 0.2 & 0.4 & 0.4 \end{bmatrix}$, and the column player's strategies are $q = \begin{bmatrix} 0.1 & 0.4 & 0.5 \end{bmatrix}$ (remember they must add up to 1). Then the average payoff for the row player is pAq.

SADDLE POINTS

In this most general version of an "m × n" game, we may not be able to determine a solution. One type of such a game that is easy to solve is a matrix game with a "saddle point." **[See "eresources: Equations and Problems" on the Website mentioned at the beginning of the chapter.]**

If a game matrix has the property that (1) a_{ij} is the minimum of the ith row, and (2) a_{ij} is the maximum of the jth column, then we call a_{ij} a *saddle point*. By inspecting A above, note that $a_{32} = 5$ is the minimum of the third row and is also the maximum of the second column. In the aforementioned case, the row player can play the third strategy and at least win \$5; the column player, playing the second strategy, minimizes his or her losses at the same \$5. Thus, in the existence of a saddle point, the value of the game, V, is the value of the saddle point. **[See "eresources: Equations and Problems" on the Website mentioned at the beginning of the chapter.]**

Consequently, the strategy vector for the row player is $p = \begin{bmatrix} 0 & 0 & 1 & 0 \end{bmatrix}$ and for the column player $q = \begin{bmatrix} 0 & 1 & 0 & 0 \end{bmatrix}$.

With simple games as described earlier, we can check for saddle points by examining each entry of the matrix. However, even in two-person zero-sum games, this can be quite complex. Consider the example of what is usually called "straight poker," where two players receive five cards from a dealer and then bet on their best hand. The number of rows and columns in such a game is 311,875,200.

SOLUTION OF ALL 2 × 2 GAMES

The general solution of such a game can be found by this two-step strategy: (1) test for a saddle point; (2) if there is a saddle point, it constitutes the solution. If there is no saddle point, solve by finding the equalizing strategy.

Describe the general 2 × 2 game as $A = \begin{vmatrix} a & b \\ d & c \end{vmatrix}$.

We proceed by assuming the row player chooses among his or her two choices with probability p; in other words, $\begin{bmatrix} p & 1-p \end{bmatrix}$. Then, as before, we find the row player's average return when the column player uses column 1 or 2. If the column player chooses the first column with probability q, his or her average losses are given when the row player uses rows 1 and 2, that is $\begin{bmatrix} q & 1-q \end{bmatrix}$.

With no saddle point, (a − b) and (c − d) are either both positive or both negative, so p is strictly between 0 and 1. If a ≥ b, then b < c; otherwise, b is a saddle point. Since b < c, we must have c > d, as otherwise c is a saddle point. Similarly, we can see that d < a and a > b. This leads us to: if a ≥ b, then a > b < c > d < a. Using an argument by symmetry, if a <= b, then a < b > c < d > a. And therefore:

If there is no saddle point, then either a > b, b < c, c > d and d < a, or a < b, b < c, c < d and d > a.

If the row player chooses the first row with probability p, the average return when the column player uses columns 1 and 2 is: ap + d(1 − p) = bp + c(1 − p). **[See "eresources: Equations and Problems" on the Website mentioned at the beginning of the chapter.]**

DOMINATED STRATEGIES

It is often possible to reduce the size (i.e., the dimensions of the game matrix) by deleting rows or columns that are clearly losing strategies for the player who might choose the row or column.

Definition: We say one row of a game matrix dominates another if every value in the first row is greater than the corresponding value in the other row. Similarly, we say that a column of the matrix dominates another if every value in the first column is less than the corresponding value in the second column.

Anything the row player can achieve by using a dominated row can be achieved at least as well by using the row that dominates. As such, a row player would never choose to play the dominated row, and a dominated row may be deleted from the matrix. Similarly, a column player can delete a dominated column from the matrix. In other words, the value of the game remains the same after dominated rows or columns are deleted. **[See "eresources: Equations and Problems" on the Website mentioned at the beginning of the chapter.]**

By a similar principle, the first row is strictly dominated by the third row, so the first row can be essentially eliminated. Thus, we are reduced to $A'' = \begin{bmatrix} 3 & 4 \\ 9 & 2 \end{bmatrix}$, and the resulting 2×2 matrix may be solved by the methods in the previous section. So, p, q, and v would be:

$$p = (c-d)/((a-b)+(c-d)) = (2-9)/((3-4)+(2-9)) = 7/8 = 0.875;$$

$$q = (c-b)/((a-b)+(c-d)) = (2-4)/((3-4)+(2-9)) = 1/4 = 0.25;$$

$$v = (ac-bd)/((a-b)+(c-d)) = (6-36)/(-1-7) = 30/8 = 3.75.$$

GRAPHICAL SOLUTIONS: 2 × n AND m × 2 GAMES

In a special case where there are either only two strategies for the row player (called a $2 \times n$ game) or only two strategies for the column player (called an $m \times 2$ game), there is an approach that can be used to solve the game using a graphical representation. With the following $2 \times n$ game, we can proceed as follows. Clearly, the row player chooses the first row with probability p; then, that player's strategy for the choice of second row must be $1 - p$. With the aforementioned example, we can calculate the average payoff for the column player with each of the n strategies available to him or her. Those payoffs are described as follows:

We can plot the potential values for the game outcome by graphing the straight line for the values of p between 0 and 1.

(A student familiar with the subject of linear programming will recognize the solution method as similar to the simplex method.)

Consider the following example for a two-row, four-column game (in other words, a $2 \times n$ game):

$$
\begin{array}{ccccc}
p & 6 & 8 & 2 & 9 \\
1-p & 8 & 4 & 10 & 2
\end{array}
$$

If the column player chooses the first column, the average payoff for the row player is $6p + 8(1 - p)$. By the same token, choices of the second, third, and fourth columns would lead to average payoffs of (column 2) $8p + 4(1 - p)$; (column 3) $2p + 10(1 - p)$; (column 4) $9p + 2(1 - p)$. Next, graph all four of these linear functions for p going from 0 to 1 (Figure 9.1).

For any fixed value of p, the row player is sure that his or her average winnings are at least the minimum of these four functions evaluated for the chosen value of p. Thus, in this range for p, we want

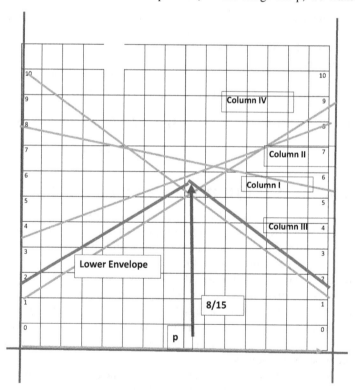

Figure 9.1 Solving the $2 \times n$ matrix game.

to find the p that achieves the maximum of this "lower envelope." In geometry or linear programming, this lower envelope is also referred to as the convex hull. In this example, we can see that maximal value for the lower envelope occurs when p = 8/15. Because the only columns that intersect at the critical point are columns 3 and 4, we can also conclude that we can reduce the game to the second and third columns, therefore once again reducing us to the 2 × 2 game for which we already have a methodology for a solution.

We can verify our graphical solution by using the algebraic approach for the solution of the reduced 2 × 2 game.

A similar approach can be used for an m × 2 game.

In theory, we can extend this graphical method to games that are 3 × n or m × 3. However, for the example in which the game has three rows instead of two, the choices of probability would yield to independent variables, say p1 and p2, and then the probabilities by row would be written as p1, p2, and 1 − (p1 + p2). Thus to model the game, we would need a three-dimensional model, where the possibilities for each variable would define a plane in a three-dimensional region.

USING GAME THEORY TO CHOOSE A STRATEGY IN THE SONY/NORTH KOREA CASE

Recall the case of the Sony Pictures hack from Chapter 5 (Sony, 2018). We consider the potential attackers we discussed as all being coordinated by some criminal mastermind, whom we will abbreviate CM. CM will play the game by choosing one of the potential attackers we considered the real attacker. How CM makes this choice will be his or her strategy. CM will be along the rows.

The cybersecurity expert (you) will be abbreviated CE and will try to determine the motivation for the attack. CE will be listed across the columns:

	CE			
CM	**Politics**	**Warfare**	**Reputation**	**Money**
North Korea				
Guardians of Peace				
WikiLeaks				
Industrial Competitor				

We assume that we know the payoffs for each possibility, as described in the following matrix. The game will be to find the optimal strategy (best probability). The payoff matrix looks like:

CM	CE			
	Politics	Warfare	Reputation	Money
North Korea	1	−1	−1	−1
Guardians of Peace	−1	−1	1	−1
WikiLeaks	−1	1	−1	−1
Industrial Competitor	−1	−1	−1	1

The rules of play are that if both CM and CE choose (Attacker, Motivation) = (North Korea, Politics), then CM wins (one unit of something).

CM also wins if neither chooses (North Korea, Politics) and CE guesses differently from CM. CE wins otherwise.

Using pure strategies, neither player can guarantee a win. The pure strategy security level for each player is therefore −1. Figuring out the mixed security strategy level "by hand" for a game with four strategies for each player is tedious and hard. In this example, three of the strategies are symmetric. This suggests a simplification.

Assume that CM always plays Guardians of Peace, WikiLeaks, Industrial Competitor, with the same probability. (This probability can be any number between 0 and 1/3; if the probability is zero, then CM always plays North Korea; if the probability is 1/3, then CM never plays North Korea and plays each of the remaining attackers with the same probability. You can view the game as heading to pure strategies for CM (either he plays Sony or he does not.) The payoff matrix becomes:

	Politics	Warfare	Reputation	Money
North Korea	1	−1	−1	−1
Not North Korea	−1	1/3	1/3	1/3

If you impose the same symmetry condition on CE, the game reduces to:

	Politics	Not Politics
North Korea	1	−1
Not North Korea	−1	1/3

It is now possible to find the mixed-strategy equilibrium for this 2×2 (i.e., two strategies for each player) game. CM's strategy will equalize the payoff she gets from either strategy choice of CE. That is, the probability of (North Korea, Politics), call it a, will satisfy a + $(1 - a)(-1) = a(-1) + (1 - a)(1/3)$.

The solution to this equation is a = 2/5. When CM mixes between North Korea and Not North Korea with probabilities 2/5 and 3/5, he gets a payoff of −1/5 whatever CE does. Similarly, you can solve for CE's strategy by equalizing CM's payoff. If b is the probability that CE plays (North Korea, Politics), then b should satisfy

$$b + (1 - b)(-1) = b(-1) + (1 - b)(1/3) \text{ or } b = 2/5.$$

Now you can go back and check that the symmetry assumption that I imposed is really appropriate. Both players are playing (North Korea, Politics) with probability 2/5 and other strategies with total probability 3/5. That means that they play each of the non-(North Korea, Politics) strategies with probability 1/3. Under this, CM expects to earn −1/5 from each of his original pure strategies, and CE can hold CM to this amount by playing (North Korea, Politics) with probability 2/5 and the other three strategies with probability of 1/5 each.

Problems can be found at https://www.routledge.com/Behavioral-Cybersecurity-Fundamental-Principles-and-Applications-of-Personality/Patterson-Winston-Proctor/p/book/9780367509798.

REFERENCES

Morris, P. 1994. *Introduction to Game Theory*. Springer, New York.
Sony Pictures Digital Productions Inc. 2018. http://www.sonypictures.com/.
Von Neumann, J., & Morgenstern, O. 1944. *Theory of Games and Economic Behavior*. Princeton University Press, Princeton, NJ.

10

THE PSYCHOLOGY OF GENDER

Since the founding of the field of psychology, there has been interest in understanding the behavior of women and men. Many popular historical and contemporary psychological theories explain patterns in the behavior of men and women. Sigmund Freud developed theories about unconscious drives that became popular in the early twentieth century. However, his theories were largely criticized for their lack of solid evidence. David Buss popularized theories of human evolution focused on mate selection that emerged in the 1990s. There have also been cross-cultural theories of gender that have been developed by psychologists to explain differences between men and women based on the characteristic of the culture. More specifically, a large body of research demonstrates that gender differences in personality traits and well-being are larger in cultures with more egalitarian gender roles, gender socialization, and sociopolitical gender equity (see Schmitt et al., 2017).

Arising in the 1960s, feminist psychologists challenged the nature of gender that was popular prior to this time (Shields & Dicicco, 2011). Thus, they began to shift approaches to understanding the psychology of gender through a move from questions centered on sex-related differences in psychological processes and outcomes to a more contextual view of gender (e.g., Deaux & Major, 1987; Marecek, 2001; Spence et al., 1975).

The extension of this contextual view of gender during the 1970s by feminist psychologists and other researchers stimulated the need for explicit differentiation "between sex as categorization on the basis of anatomy and physiology, and gender as a culturally defined set of meanings attached to sex and sex difference" (Shields & Dicicco, 2011, p. 493).

Research on prenatal and postnatal gender development had a profound influence on this shift in differentiation between sex and gender psychological conceptualizations (e.g., Spence & Helmreich, 1978). Following these developments in psychology, there was a wave of pioneering research that adopted a more social contextual view of the psychology of gender (e.g., Deaux & Major, 1987; Eagly & Wood, 1999; Eccles, 1987). And this is a view of gender that persists today among psychologists, though some psychologists have argued that more complex formulations of gender differences are necessary (Fine & Gordon, 1991; Shields & Bhatia, 2009).

DEFINITIONS AND ANALYSIS OF GENDER

Within the field of psychology, gender has been defined and analyzed in a variety of ways. Stewart and McDermott (2004) identified the following dominant orientations psychologists use to defining gender: (1) gender as sex differences on outcomes, (2) gender as a role and gendered socialization, (3) gendered power relations, and (4) intersections of gender identity with other social and sociostructural identities (e.g., race, ethnicity, sexual orientation).

GENDER-AS-TRAIT: THE SEX DIFFERENCES APPROACH

Within this approach to gender psychology, the central question that psychologists pursue is: How and why do average differences in attitudes, ability, personality traits, and other behavioral tendencies appear? This approach assumes that differences arise from preexisting "essential" differences between males and females. The core idea of this approach is that differences between male and females are "natural, deep-seated, and of profound personal and social consequences. This proposal easily built upon Anglo-American belief in 'natural' gender differences as differentiating 'advanced' races from more primitive" (Shields & Dicicco, 2011, p. 491). In essence, this approach explains the differences between males and females as a result of genes and hormones (Kitzinger, 1994).

GENDER IN SOCIAL CONTEXT: THE WITHIN GENDER VARIABILITY APPROACH

In this approach to gender psychology, the central question that psychologists pursue is: Within highly gendered psychological phenomena, what are the sources of within-gender variation? Highly gendered psychological phenomena are defined by average differences that researchers have discovered between men and women in various studies (i.e., attitudes, ability, motivation, personality traits, and other behavioral tendencies). More specifically, psychologists have demonstrated how the social environment shapes individuals' expectation of success, ideas about the importance of a task, and perception of available options, as well as academic and career choices in science and mathematics (e.g., Eccles et al., 1983; Eccles, 2011).

GENDER LINKED TO POWER RELATIONS APPROACH

In this approach to gender psychology, the central question that psychologists pursue is: How does gender structure social institutions' practices, norms, and policies within which men and women operate? With this approach, gender beliefs and behaviors are ideologies that are embedded in sociostructural systems (Shields & Dicicco, 2011). Within this approach, "gender" does not merely operate at the level of sex differences or as the result of social interactions in which beliefs about gender are expressed in actions that actually create confirming evidence for those beliefs. Instead, "gender" also operates in the social structures that define power relationships throughout culture (Stewart & McDermott, 2004, p. 521). Psychologists have explored leadership, marital relationships, decision-making (i.e., choice) and conflict, and task performance using this conceptualization of gender psychology. There are various configurations of relationships in which these behaviors take place, including dyads, organizational hierarchies, and broader cultural political structures (Deaux & Major, 1987).

GENDER AS INTERSECTIONAL: THE IDENTITY ROLE, SOCIAL IDENTITY, AND SOCIAL STRUCTURAL APPROACH

Within this approach to gender psychology, the central question that psychologists pursue is: How do gender roles, social identities, and social structural dynamics operate individually and interactionally to shape psychological processes and outcomes of men and women? This approach adopts orientations grounded in personal individual identity theory rooted primarily in personality psychology; social identity theory anchored largely within social identity theory; racial identity theory cutting across personality, social, and developmental psychology; and intersectionality theory that encompasses boundaryless subareas of psychology. The kinds of topics that psychologists have explored within this approach are ego identity development, gender role identity (Eccles, 1987, 2009), social identity (Gurin, 1985), racial and ethnic identity (e.g., Sellers et al., 1998), and intersectionality (e.g., Cole, 2009; Crenshaw, 1994; Hurtado & Sinha, 2008; Ireland et al., 2018). One example of gender role theory was the research of psychologists who explored gender socialization in terms of the experience of having one's behavior, beliefs, and attitudes shaped by culturally defined, gender-specific roles (Shields & Dicicco, 2011). For example, increasingly, psychologists are adopting feminist theories and critical race theories to develop theoretical, methodological, and practical formulations of intersectionality, which refers to the simultaneous meaning and consequences of multiple categories of identity, difference, and disadvantage, particularly related to the intersections of race, gender, and social class (Cole, 2009).

THE NATURE VERSUS NURTURE DEBATE IN GENDER PSYCHOLOGY

One common way to classify these dominant approaches to the psychology of gender is as a result of nature, nature, or a combination of the two. This classification has been hotly debated by psychologists for decades. The fundamental question that undergirds this debate is as follows: Is nature or nature responsible for differences and similarities found in the beliefs, attitudes, abilities, motives, personality traits, and other behavioral tendencies of men and women? It is common for researchers to focus on one cause to the exclusion of the other or to treat them as competing explanations (see Eagly & Wood, 2013).

CONCLUSION

Psychologists have used a variety of definitions of gender across time. There are promising are new areas of research on the brain and behavior that are opening up new questions and providing insights about the psychology of gender. Understanding the varied approaches psychologists use to understand gender can inform how to identify, analyze, and manage behavior involved in cybersecurity events faced by individuals, as well as within organizations.

REFERENCES

Cole, E. R. 2009. Intersectionality and research in psychology. *American Psychologist*, 64(3), 170–180.

Crenshaw, K. W. 1994. Mapping the margins: Intersectionality, identity politics, and violence against women of color. In M. A. Fineman and R. Mykitiuk (Eds.), *The Public Nature of Private Violence*. Routledge, New York, pp. 93–118.

Deaux, K., & Major, B. 1987. Putting gender into context: An interactive model of gender-related behavior. *Psychological Review*, 94(3), 369–389.

Eagly, A. H., & Wood, W. 2013. The nature–nurture debates: 25 Years of challenges in understanding the psychology of gender. *Perspectives on Psychological Science*, 8, 340–357.

Eagly, A., & Wood, W. 1999. The origins of sex differences in human behavior: Evolved dispositions versus social roles. *American Psychologist*, 54, 408–423.

Eccles, J. S. 1987. Gender roles and women's achievement-related decisions. *Psychology of Women Quarterly*, 11(2), 135–172.

Eccles, J. S. 2009. Who am i and what am i going to do with my life? Personal and collective identities as motivators of action. *Educational Psychologist*, 44(2), 78–89.

Eccles, J. S. 2011. Understanding women's achievement choices: Looking back and looking forward. *Psychology of Women Quarterly*, 35(3), 510–516.

Eccles J. S., Adler, T. F., Futterman, R., Goff, S. B., Kaczala, C. M., Meece, J. L., & Midgley, C. 1983. Expectancies, values and academic behaviors. In J. Spence (Ed.), *Achievement and Achievement Motivation*. W.H. Freeman and Co., San Francisco, CA, pp. 75–146.

Fine, M., & Gordon, S. M. 1991. Effacing the center and the margins: Life at the intersection of psychology and feminism. *Feminism and Psychology*, 1(1), 19–27.

Gurin, P. 1985. Women's gender consciousness. *Public Opinion Quarterly*, 49(2), 143–163.

Hurtado, A., & Sinha, M. 2008. More than men: Latino feminist masculinities and intersectionality. *Sex Roles*, 59(5–6), 337–349.

Ireland, D., Freeman, K. E., Winston-Proctor, C. E., DeLaine, K. D., Lowe, S. M., & Woodson, K. M. 2018. (Un)Hidden figures: A synthesis of research addressing the intersectional experiences of Black women and girls in STEM education. *Review of Research in Education*, 42(1), 226–254.

Kitzinger, C. 1994. Should psychologists study sex differences? *Feminism and Psychology*, 4(4), 501–506.

Marecek, J. 2001. After the facts: Psychology and the study of gender. *Canadian Psychology/Psychologie Canadienne*, 42(4), 254–267.

Schmitt D. P., Long A.E. McPhearson A., O'Brien K., Remmert B., & Shah S.H. 2017. Personality and gender differences in global perspective. *International Journal of Psychology*, 52(1), 45–56.

Sellers, R. M., Smith, M. A., Shelton, J.N., Rowley, S. A. J., & Chavous, T. M. 1998. Multidimensional model of racial identity: A reconceptualization of African American racial identity. *Personality and Social Psychological Review*, 2, 18–39.

Shields, S. A., & Bhatia, S. 2009. Darwin and race, gender, and culture. *American Psychologist*, 64, 111–119.

Shields, S. A., & Dicicco, E. C. 2011. The social psychology of sex and gender: From gender differences to doing gender. *Psychology of Women Quarterly*, 35(3), 491–499.

Spence, J. T., & Helmreich, R. L. 1978, *Masculinity and Femininity: Their Psychological Dimensions, Correlates, and Antecedents.* University of Texas Press, Austin.

Spence, J. T., Helmreich, R., & Stapp, J. 1975. Ratings of self and peers on sex-role attributes and their relations to self-esteem and conceptions of masculinity and femininity. *Journal of Personality and Social Psychology*, 32, 29–39.

Stewart, A., & McDermott, C. 2004. Gender in psychology. *Annual Review of Psychology*, 55, 519–544.

11

TURING TESTS

This chapter will consider both the history of the human/machine Turing test and, for the purposes of our understanding of behavioral cybersecurity, report on actual results carried out in our courses regarding the "gender Turing test" and also discuss a similar concept the "age Turing test."

A theme that we have continued to raise throughout this book involves the potential tools that a defender of a cyberattack might have at his or her disposal in order to defeat the attack in question.

One area for examination is the question of whether a suspected cyberattack comes from a human or machine opponent. In fact, this question has been at the forefront of computer science almost from its beginnings and indeed will likely grow in its importance in the future.

Alan Turing first proposed a test, which, if it could be answered, could distinguish a third party in a dialogue as being human or a computer. He proposed an experiment with two participants in separate rooms so they had no visual contact. In one room was a questioner who could ask any question at all of the entity in the other room, who could be a human or computer. If, after the questioner was allowed as many questions as possible of the other entity, the questioner could not determine with absolute certainty if the entity was a human or a machine, and indeed it was a machine, and it could be said that the machine possessed human intelligence. Turing published this article in the journal *Mind* in 1950.

One of the earliest initiatives to test the Turing challenge was a program called Eliza developed by Joseph Weizenbaum at MIT in 1966 (Weizenbaum, 1966). With the Eliza program, a questioner could make as many statements or questions as desired of Eliza, who was allegedly a psychotherapist. Here is a short dialogue from the original Eliza program:

ELIZA: Is something troubling you?
YOU: I'm not feeling well today.

ELIZA: Do you know anyone else who is not feeling well today?
YOU: No, just me.
ELIZA: Why not?
YOU: I have a headache.
ELIZA: You say you have a headache?
YOU: Yes.
ELIZA: I understand.
YOU: What can I do to get rid of my headache.
ELIZA: Is it important to you that you have your headache?
YOU: Yes, I definitely need to end my headache.
ELIZA: You seem to be quite positive.
YOU: Absolutely.
ELIZA: Does that have anything to do with the fact that you have
your headache?

The usual result of Eliza tests was for the human user to quit in frustration at the repetitiveness of Eliza's answers. No person would reach the conclusion that Eliza was actually a human.

It is interesting that in the original formulation of the Turing test, Turing described the process with the questioner as indicated earlier, but the respondent not visible in a nearby room would be either a male or a female. Undoubtedly, Turing described the challenge in 1950 realizing that it would merely confuse his readers to imagine a computer (or an electronic machine) in another room, since at that time there were only a handful of computers in the world and very few readers would have understood the context and importance of the problem.

INTRODUCTION

In developing this field of behavioral cybersecurity, we have researched the possibility of detecting the gender of a writer, such as a hacker in a computing environment.

It has been noted in the current research that posing the Turing test challenge in the context of gender determination is in fact the manner by which Turing himself chose to explain the concept of his test to an audience that might have been challenged by the idea that machines could conduct a dialogue with human interrogatories.

THE ROLE OF THE TURING TEST IN BEHAVIORAL CYBERSECURITY

Given the confluence of external events, the power of the Internet, increasing geopolitical fears of "cyberterrorism" dating from 9/11, a greater understanding of security needs and industry, and economic projections of the enormous employment needs in cybersecurity, many universities have developed more substantial curricula in this area, and the US National Security Agency has created a process for determining Centers of Excellence in this field (NCAEIAE, 2018).

At the 1980 summer meeting of the American Mathematics Society in Ann Arbor, Michigan, a featured speaker was the distinguished mathematician the late Peter J. Hilton (Pedersen, 2011). Dr. Hilton was known widely for his research in algebraic topology, but on that occasion, he spoke publicly for the first time about his work in cryptanalysis during World War II at Hut 8 in Bletchley Park, the home of the now-famous efforts to break German encryption methods such as the Enigma.

The first author was present at that session and has often cited Professor Hilton's influence in sparking interest in what we now call cybersecurity. Hilton at the time revealed many of the techniques used at Bletchley Park in breaking the Enigma code. However, one that was most revealing was the discovery by the British team that, contrary to the protocol, German cipher operators would send the same message twice, something akin to "How's the weather today?" at the opening of an encryption session. (This discovery was represented in the recent Academy Award–nominated film *The Imitation Game* [Sony, 2014].) Hilton was portrayed in *The Imitation Game*. Of course, it is well known in cryptanalysis that having two different encryptions of the same message with different keys is an enormous clue in breaking a code. Thus, it is not an exaggeration to conclude that a behavioral weakness had enormous practical consequences, as the Bletchley Park teams have been credited with saving thousands of lives and helping end the war.

A FINAL EXAM QUESTION

In an offering of the undergraduate course Behavioral Cybersecurity at Howard University in the spring semester 2016, there was considerable discussion about the identification of the gender of a potential hacker. This led to a question on the final examination that asked students the following:

> We know, in general in the US as well as at Howard [at the time, the first author's university], that only about 20% of Computer Science majors are female. Furthermore, of those CS students choosing to concentrate in Cybersecurity, fewer than 10% are female. Can you suggest any reason or reasons that so many fewer female computer scientists choose Cybersecurity?

Readers may find the text of all students' responses to that final exam question at Patterson & Winston-Proctor (2019).

WHILE GRADING

In the course of grading this examination, the first author, as he read each answer, questioned himself as to whether he could determine if the author of the answer was one of his male or female students, based on the tone, language choice, use of certain keywords, and expected perception of the point of view of the author. Thus, as he found that this was very often difficult to determine, it seemed that it might be interesting for other persons of widely varied backgrounds—for example, gender, age, profession, geographic location, first language—to be posed the same questions.

Consequently, a test was constructed from the student responses. Three variations were added: The first author himself wrote one of the responses in an attempt to deceive readers into thinking the writer was female, and two responses were repeated in order to validate whether the responders could detect the repetition, thus showing that they were concentrating on the questions themselves.

The specific questions are listed at the end of the chapter.

TURING'S PAPER IN MIND

It was noted that there was a certain similarity between the administration of this test and the classic Turing test originally posed by Turing to respond to the proposition that machines (computers) could possess intelligence.

The Turing test supposes that a questioner is linked electronically to an entity in some other location, and the only link is electronic.

In other words, the questioner does not know whether he or she is corresponding with a human being or a computer. The test is conducted in the following way: The questioner may ask as many questions as he or she desires, and if, at the end of the session, the questioner can absolutely determine that the invisible entity is human, and in fact it is a computer, then the other entity can reasonably be said to possess intelligence.

There has been a large body of research on this topic since Turing first posed the question around 1950, beginning with the development of the artificial "psychologist" named Eliza, originally developed by Weizenbaum (1966), and leading all the way to IBM supercomputer Watson, which was able to beat two human experts on the game show Jeopardy! in 2010 (Baker, 2012). Regarding the Jeopardy experiment, Watson was the overall winner, but on some questions where there may have been semantic interpretations, Watson simply failed, whereas on many others, the speed of Watson enabled it to beat the human contestants. On a more general level, it is generally accepted, however, that no computer, however, powerful, has been able to pass the Turing test.

It has been discussed throughout the history of computer science whether this test has been satisfied or indeed if it could ever be satisfied.

THE IMITATION GAME

It seems very interesting, in the context of the gender test as described in our course, that in many ways it draws historically from Turing's thinking.

Many readers may note that the recent film, as indicated earlier, addressing both Turing's life and his efforts in breaking the Enigma Code in the World War II was called *The Imitation Game*. Turing published an extremely important article in the May 1950 issue of *Mind* entitled "Computing Machinery and Intelligence." More to the point, he called the section of this paper in which he first introduced his Turing test *The Imitation Game*, which evolved into the title of the biographical film.

It is significant, in our view, that in order to explain to a 1950s audience how to establish whether an entity possessed intelligence that to describe the test in terms of a human and machine would be incomprehensible to most of his audience, since in the 1950, there were only a handful of computers in existence.

Consequently, in introducing the nature of his test, he described it as a way of determining gender, as follows:

> I propose to consider the question, 'Can machines think?' This should begin with definitions of the meaning of the terms 'machine' and 'think'. The definitions might be framed so as to reflect so far as possible the normal use of the words, but this attitude is dangerous. If the meaning of the words 'machine' and 'think' are to be found by examining how they are commonly used it is difficult to escape the conclusion that the meaning and the answer to the question, 'Can machines think?' is to be sought in a statistical survey such as a Gallup poll. But this is absurd. Instead of attempting such a definition I shall replace the question by another, which is closely related to it and is expressed in relatively unambiguous words.
>
> The new form of the problem can be described in terms of a game which we call the 'imitation game'. It is played with three people, a man (A), a woman (B), and an interrogator (C) who may be of either sex. The interrogator stays in a room apart from the other two. The object of the game for the interrogator is to determine which of the other two the man is and which is the woman. He knows them by labels X and Y, and at the end of the game he says either 'X is A and Y is B' or 'X is B and Y is A'. The interrogator is allowed to put questions to A and B thus:
>
> C: Will X please tell me the length of his or her hair?

We thus view the initiative that we have developed from the Behavioral Cybersecurity course as a descendant in some small way of Turing's proposition.

RESPONDENTS

In order to understand the ways in which persons interpret written text and try to assign gender to the author—in effect a version of the gender Turing test (henceforth, GTT) described by Turing in the paper cited earlier—a number of individuals from varying backgrounds, genders, ages, first languages, countries, and professions were given the test in question.

There were 55 subjects completing this test, and a description of their demographics is as follows.

The participants were selected as volunteers, primarily at occasions where the first author was giving a presentation. No restrictions were placed on the selection of volunteer respondents, nor was there any effort taken to balance the participation according to any demographic objective.

The voluntary subjects were (except on one occasion) given no information about the purpose of the test and were also guaranteed anonymity in the processing of the test results. There was no limit on the time to take the test, but most observed respondents seemed to complete the test in about 15 minutes.

SUMMARY OF RESULTS

The responses were scored in two ways. First, the number of correct answers identifying the student author was divided by the total number of questions (24) in the complete test. Alternatively, the score was determined by the number of attempts. Since in only 2 of 19 instances, the difference between the two exceeded 2%, it was decided to use the second set of response scores.

Observations of the results of these responses from the 55 participants in this study and their very diverse experiences that they brought to the response to this test yield some very interesting questions to ponder:

First, female respondents were more accurate in the identification of the gender of the students by a margin of 56.89%–51.02%.

Next, older respondents were more accurate in their identification than younger responses by a similar margin of 57.6%–51.77%. This might be a more surprising result since for the most part, the older respondents were not as technically experienced in computer science or cybersecurity matters than the younger responders, who for the most part were students themselves.

One very clear difference is that Eastern European respondents scored far higher in their correct identification of the students' gender, averaging 66.67%, with the nearest other regional responses being fully 10% less. The number of respondents from Eastern Europe was very small, so generalizations might be risky in this regard. However, the Eastern Europeans (from Romania and Russia) were not first-language speakers of English, although they were also quite fluent in the English language. Each of them also tied for the highest percentage of correct answers of anyone among all 55 respondents.

Of the respondents from the various disciplines, the linguists, anthropologists, engineers, and psychologists all fared better than the computer scientists—and lowest of all were the students who took the test (as opposed to the students who wrote the original answers).

It is possible, of course, to view the entire data set of responses to this test as a matrix of dimensions 24×55, wherein the students who wrote the original exam—and thus in effect, created the GTT—represent the rows of the matrix, and the gender classifications by the 55 responders are the columns. If we instead examine the matrix in a row-wise fashion, we learn of the writing styles of the original test takers and their ability (although inadvertent, because no one, other than the first author, planned that the writings would be used to identify the gender of the writer).

Thus, it is perhaps more informative than the assessment of the ability of the respondent to determine the gender of the test takers to note that several of the original test takers were able, unconsciously, to deceive over two-thirds of the respondents. Fully one-quarter (6 of 24) of the students reached the level of greater than two-thirds deception. Of these six "high deceivers," three were female and three were male students.

At the other end of the spectrum, one-third of the students were not very capable of deception—fooling less than one-third of the respondents. Of these eight students, six were male and only two were female. On the whole, averaging the level of deception by the male and female students, on average the female students were able to deceive 52.5% of the respondents, whereas the male students were only able to accomplish this with 42.2% of the respondents.

"COACHING" RESPONDENTS

All of the respondents described earlier had simply been given a test with only the simple instruction described in the attachment, without any prior preparation or understanding on the part of the respondent as to possible techniques for identifying the gender of a writer or author.

Consequently, we determined that it would be useful to see if persons could be given some training in order to try to improve their results on the GTT. We attempted to identify a number of keys that would assist a reader in trying to improve their scores on the GTT or related tests.

Our next objective was to see if a subject could improve on such text analysis in the case of distinguishing the gender of a writer by looking for certain clues that could be described. A number of techniques to identify the gender of an author were described to perform an analysis of the questions in the original GTT:

1. Examine how many pronouns are being used. Female writers tend to use more pronouns (I, you, she, their, myself).
2. What types of noun modifiers are being used by the author? Types of noun modifiers: a noun can modify another noun by coming immediately before the noun that follows it. Males prefer words that identify or determine nouns (a, the, that) and words that quantify them (one, two, more).
3. Subject matter/style: the topic dealt with or the subject represented in a debate, exposition, or work of art. "Women have a more interactive style," according to Shlomo Argamon, a computer scientist at the Illinois Institute of Technology in Chicago.
4. Be cognizant of word usage and how it may reveal gender. Some possible feminine keywords include with, if, not, where, be, should. Some of the other masculine keywords include around, what, are, as, it, said. This suggests that language tends to encode gender in very subtle ways.
5. "Women tend to have a more interactive style," said Shlomo Argamon, a computer scientist at the Illinois Institute of Technology in Chicago (Argamon et al., 2003). "They want to create a relationship between the writer and the reader."

    Men, on the other hand, use more numbers, adjectives, and determiners—words such as "the," "this," and "that"—because they apparently care more than women do about conveying specific information.
6. Pay attention to the way they reference the gender of which they speak. For example, a female may refer to her own gender by saying "woman" rather than "girl."
7. Look at the examples that they give. Would you see a male or female saying this phrase?
8. A male is more likely to use an example that describes how a male feels.
9. Women tend to use better grammar and better sentence structure than males.
10. When a person of one gender is describing the feelings/thoughts of the opposite gender, they tend to draw conclusions that make sense to them but will not provide actual data.

It should be noted that some prior work includes the development of an application available on the Internet (Gender Guesser), developed by Neil Krawetz based on Krawetz (2018) and described at the location http://hackerfactor.com/GenderGuesser.php.

This application seems to depend on the length of the text being analyzed and, in comparison with the responses of our human responders, does not perform as well, as normally the application indicates that the text is too short to give a successful determination of gender.

However, because the overall objective of this research is to determine if a GTT can be used in a cybersecurity context, it is likely that an attacker or hacker might only be providing very short messages—for example, a troll on the Internet trying to mask his or her identity in order to build a relationship, say, with an underage potential victim.

FUTURE RESEARCH

The questions that have been raised by this research have also opened the potential of devising other such tests to determine other characteristics of an author, such as age, profession, geographic origin, or first language. In addition, given that the initial respondents to the test as described earlier are themselves from a wide variety of areas of expertise, nationality, and first language, a number of the prior participants have indicated interest in participating in future research in any of these aforementioned areas.

PROBLEMS

1. Read Turing's article in *Mind*. How would a reader respond (a) in 1950; (b) today?
2. Try Eliza. How many questions did you ask before Eliza repeated?
3. Construct three questions for Jeopardy! that the human contestants would probably answer faster or more correctly than Watson.
4. Give examples of two encryptions of the same message with different keys. Use the encryption method of your choice.
5. Consider developing your own gender Turing test. List five sample questions that might differentiate between a female or male respondent.
6. Consider developing an age Turing test. List five sample questions that might differentiate between a younger or older respondent.

7. Take the "Who Answered These Questions" gender Turing test. To find out your score, email your responses to waynep97@gmail.com. Send a two-line email. The first line will have "F: x1, 2, x3, …" and the second line "M: y1, y, y3, …." You will receive an email response.

8. Comment in the context of 2019, from Turing's article in *Mind*, on interrogator C's question in order to determine the gender of the hidden subject.

9. Run the Gender Guesser on (A) your own writing and (B) the "Test Bed for the Questionnaire" in the following.

REFERENCES

Argamon, S., Koppel, M., Fine, J., & Shimoni, A. R. 2003. Gender, genre, and writing style in formal written texts. *Text*, 23(3), 321–346.

Baker, S. 2012. *Final Jeopardy: The Story of Watson, the Computer That Will Transform Our World*. Houghton Mifflin Harcourt, Boston, MA.

Krawetz, N. 2018. Gender Guesser. http://hackerfactor.com/Gender Guesser.php.

National Centers of Academic Excellence in Information Assurance Education (NCAEIAE). 2018. National Security Agency. https://www.nsa.gov/ia/academic_outreach/nat_cae/index.shtml.

Patterson, W. & Winston-Proctor, C. 2019. *Behavioral Cybersecurity*. CRC Press, Boca Raton, FL.

Pedersen, J. (ed.). 2011. Peter Hilton: Codebreaker and mathematician (1923–2010). *Notices of the American Mathematics Society*, 58(11), 1538–1551.

Sony Pictures Releasing. 2014. *The Imitation Game* (film).

Turing, A. M. 1950. Computing machinery and intelligence. *Mind: A Quarterly Review of Psychology and Philosophy*, LIX(236), 433–460.

Weizenbaum, J. 1966. ELIZA—A computer program for the study of natural language communication between man and machine. *Communications of the Association for Computing Machinery*, 9, 36–45.

12

MODULAR ARITHMETIC AND OTHER COMPUTATIONAL METHODS

Please see edition equations online at https://www.routledge.com/Behavioral-Cybersecurity-Fundamental-Principles-and-Applications-of-Personality/Patterson-Winston-Proctor/p/book/9780367509798.

Critical to the understanding of almost all cryptographic methods is an understanding of the study of what are called "modular arithmetic systems."

Any cryptographic method must ensure that when we perform the encryption step, we must be able to arrive back at the original or plaintext message when we perform the decryption. As an immediate consequence, this implies that representing information in computer memory cannot be interpreted as in the computer context as floating point numbers, or, in mathematical terminology, real or rational numbers. The problem that arises is that interpretation of the result of any computation in computer memory of real or rational numbers leads to an uncertainty in terms of the lowest-order bit or bits of that computation. Therefore, in general, if we were to manipulate real or rational numbers in a cryptographic computation, the fact that the lowest-order bits are indeterminate could mean that no true or exact inversion could be performed.

As a consequence, virtually all cryptosystems use the natural numbers or integers as the basis for computation. Indeed, since computations in the integers might apply an unlimited range, instead we almost always use a smaller, finite, and fully enumerable system derived from the integers that we generally refer to as "modular arithmetic."

Z$_n$ OR ARITHMETIC MODULO n

A standard definition of the integers can be written as $\mathbf{Z} = \{\ ,\ ...,$ $-2, -1, 0, 1, 2, ...,\ \}$ with operations $+, \times$. Of course, this set $\mathbf{Z}$ is infinite, so we derive a finite variant of the integers that we refer to as the "integers modulo n," written $\mathbf{Z}_n$ and defined as

$$\mathbf{Z}_n = \{0, 1, 2, ..., n-1\};$$

and if a and b $\in \mathbf{Z}_n$, a + b is defined as the remainder of a + b when divided by n, and a $\times$ b is defined as the remainder of a $\times$ b when divided by n.

A set of elements with a binary operation (such as $\mathbf{Z}_n$ with +) forms a group G = {a, b, ...} if several conditions are satisfied:

1. *Closure*: If a, b $\in$ G, so is a + b.
2. *Identity*: There is a special element called the identity, i, such that a + i = i + a = a, for all a in G.
3. *Associativity*: For all a, b, c $\in$ G, a + (b + c) = (a + b) + c.
4. *Inverse*: For all a, there exists some b (that we write b = −a) such that a + b = a + (−a) = i.

In the case of $\mathbf{Z}_n$ with the + operation, the identity is 0, and all four conditions are satisfied, so $\mathbf{Z}_n$ with addition forms a group.

In the case of $\mathbf{Z}_n$ with the + operation, the identity is 0, and all four conditions are satisfied, so $\mathbf{Z}_n$ with addition forms a group.

Let us take an example, first $\mathbf{Z}_6$:

Addition in $\mathbf{Z}_6$

+	0	1	2	3	4	5
0	0	1	2	3	4	5
1	1	2	3	4	5	0
2	2	3	4	5	0	1
3	3	4	5	0	1	2
4	4	5	0	1	2	3
5	5	0	1	2	3	4

Multiplication in $\mathbf{Z}_6$

×	0	1	2	3	4	5
0	0	0	0	0	0	0
1	0	1	2	3	4	5
2	0	2	4	0	2	4
3	0	3	0	3	0	3
4	0	4	2	0	4	2
5	0	5	4	3	2	1

Now consider $\mathbf{Z}_7$:
Addition

+	0	1	2	3	4	5	6
0	0	1	2	3	4	5	6
1	1	2	3	4	5	6	0
2	2	3	4	5	6	0	1
3	3	4	5	6	0	1	2
4	4	5	6	0	1	2	3
5	5	6	0	1	2	3	4
6	6	0	1	2	3	4	5

Multiplication

×	0	1	2	3	4	5	6
0	0	0	0	0	0	0	0
1	0	1	2	3	4	5	6
2	0	2	4	6	1	3	5
3	0	3	6	2	5	1	4
4	0	4	1	5	2	6	3
5	0	5	3	1	6	4	2
6	0	6	5	4	3	2	1

WHAT ARE THE DIFFERENCES IN THE TABLES?

An element (in any multiplication table) has a multiplicative inverse if there is a 1 in the row corresponding to that element. Elements with multiplicative inverses in Z_6 are 1 and 5; all nonzero elements in Z_7 have inverses. The essential difference between 6 and 7 is that 7 is prime and 6 is not.

In general, in a modular arithmetic system based on the number n, if n is a composite number, there will always be some pair of numbers a and b less than n whose product will be 0 in the multiplication table, and in this case, neither a nor b will have an inverse in the multiplication operation in Z_n; if, however, the modular system is based on a prime number—let us call it p—then every nonzero element in Z_p will have a multiplicative inverse.

The reason, of course, is that in the case that n is composite, if you take two factors of n, a and b, greater than 1, then $a \times b = n$, that is, $a \times b \equiv 0 \pmod n$; then you can see from the table that neither a nor b can have a multiplicative inverse.

The result of these observations is that in the case of a prime number p, every nonzero element in the system has an inverse under multiplication. Therefore, the Z_p system (omitting the zero in the case of multiplication) contains two group structures, one for addition and one for multiplication. If we can add one other condition (which is indeed satisfied for all n whether prime or composite) called the distributive law relating addition and multiplication, that is $a \times (b + c) = a \times b + a \times c$ for all a, b, and c, then the overall modular system is considered a mathematical field.

Therefore, we can conclude that the modular systems Z_p are fields when p is prime, and Z_n are not fields when n is composite.

Prime numbers, again, are those with no proper divisors, for example, 2, 3, 5, ..., 13, 17, ..., 23, 29, ...

For any natural number n, call $\phi(n)$ the Möbius ϕ-function. It counts how many numbers between 1 and n are relatively prime to n—that is, have no common factors greater than 1. Clearly, if the number n is a prime, it has no factors greater than 1, so $\phi(p) = (p - 1)$.

In general, for large numbers n, $\phi(n)$ is infeasible to compute. We know that if the number is a prime, p, then $\phi(p) = (p - 1)$. Also, if n is the product of only two primes p and q ($n = pq$), then $\phi(n) = (p - 1) \times (q - 1)$.

There is one extremely important result about the Möbius function that arises many times in cryptography. We will state this without proof, but that can be found in any elementary college algebra book.

For any n, if you construct the mod n system, and for any $a < n$, that is relatively prime to n (alternatively, the greatest common divisor or GCD, of a and n is $GCD(a, n) = 1$). Then in this case, raising a to the $\phi(n)$ power gives: power gives $a^{\phi(n)} (\text{mod } n) = 1$.

This is sometimes called the "little Fermat theorem." (Please see online.)

FINITE FIELDS

This system can also be thought of as the integers $\mathbf{Z}$, $\mathbf{Z}/(p)$, which means in this new system, we collapse all the values that have the same remainder mod p.

We saw that if p is a prime, the system $\mathbf{Z}_p$ has the special property that all nonzero elements have multiplicative inverses; that is, for any $a \neq 0$, there exists some b for which $a \times b \equiv 1 \ (\text{mod } p)$. This system can also be thought of as the integers $\mathbf{Z}$, $\mathbf{Z}/(p)$, which means in this new system, we collapse all the values that have the same remainder mod p.

Such an algebraic system mod p with addition and multiplication is called a *field*. In fact, such a (finite) field can be defined for all prime numbers p.

We can go a little further with finite fields. We can define the system of all polynomials $\mathbf{Z}_p[x]$ in a single variable, then $\mathbf{Z}_p[x]/(q(x))$, where $q(x)$ is an *irreducible polynomial* of degree n.

The addition and multiplication of polynomials is as usual, except that the coefficients of the polynomials in the system are always modulo p.

So, for example, in the modulo 13 system of polynomials $\mathbf{Z}_{13}[x]$, if we have $p1 = (3x^2 + 4x + 2)$ and $p2 = (x^3 + 5x^2 + 10x + 5)$, then

$$p1 + p2 = \left(3x^2 + 4x + 2\right) + \left(x^3 + 5x^2 + 10x + 5\right) = \left(x^3 + 8x^2 + 14x + 7\right)$$

$$= \left(x^3 + 8x^2 + x + 7\right)(\text{mod } 13);$$

$$p1 \times p2 = \left(3x^2 + 4x + 2\right) \times \left(x^3 + 5x^2 + 10x + 5\right) = 3x^5 + (15 + 4)x^4$$

$$+ (1 + 20 + 30)x^3 + (15 + 40 + 10)x^2 + (20 + 20)x + 10$$

$$= 3x^5 + 19x^4 + 51x^3 + 65x^2 + 40x + 10$$

$$= 3x^5 + 6x^4 + 12x^3 + 0x^2 + x + 10(\text{mod } 13).$$

Irreducible polynomials q(x) are like prime numbers—they cannot be factored (beyond factoring coefficients). By analogy, the system where we collapse polynomials with the same remainder mod q(x) also becomes a field, which we call GF(p, n), the Galois field. Again, p is the mod system for the coefficients, and n indicates the degree of the polynomials—once we divide by q(x), no term will remain with an exponent higher than $(n - 1)$.

Of the GF(p, n), several of the form GF(2,n) are the key components of the current US government standard for data encryption, known as the advanced encryption standard (AES) or Rijndael.

THE MAIN RESULT CONCERNING GALOIS FIELDS

The theory of these systems was a result developed by Evariste Galois, in the early nineteenth century, stating that all algebraic *fields* with a finite number of elements can be described as a GF(p, n) (including the fields Z_p, since they can be thought of as GF(p, 1), that is, dividing by an irreducible polynomial of degree 1 [such as ax + b]).

Furthermore, all of the possible choices for a Galois field of type GF(p, n) are equivalent, and their number of elements is p^n.

And a bit about Galois himself: He lived in the early nineteenth century in Paris. He developed these very important results in algebra while a teenager. He was also a political radical and went to prison. Upon his release, his interest in a young woman led to a duel in which he was killed at age 21.

MATRIX ALGEBRA OR LINEAR ALGEBRA

A matrix or array is a set of numbers of some type (integers, rational, or real, for example) considered as a rectangular set with a certain number of rows or columns. We say that a matrix is of order m × n if it has m rows and n columns and consequently has m × n elements all together.

Here is an example of a 4 × 3 matrix A with real number values. We usually write a matrix enclosing its values in square brackets. **[See "eresources: Equations and Problems" on the Website mentioned at the beginning of the chapter.]**

The usual compact notation for A is $[a_{ij}]$, where i and j enumerate the elements in the rows and columns, respectively.

Under certain conditions, the operations of the system of the matrix elements can be extended to an operation on the matrices themselves. First, regarding addition, two matrices can only be added if they have the same dimension, m × n. **[See "eresources: Equations and Problems" on the Website mentioned at the beginning of the chapter.]**

Matrices can also be multiplied. The conditions for being able to do this are if you have A (with m rows and n columns) and B (with n rows and p columns), then A and B can be multiplied to form a matrix C, with C having m rows and p columns. **[See "eresources: Equations and Problems" on the Website mentioned at the beginning of the chapter.]**

Why is this useful? On the one hand, it provides a useful and compact way of describing an array of values. But on the other hand, perhaps the best example of this notation is how we can use it to translate a system of linear equations into a single matrix equation.

In the special case where you have a set of n linear equations in n unknowns, replacing the set of equations by a single matrix equation AX = B leads to a method of solving the entire system by solving the one matrix equation. Necessarily, because there are n equations in n unknowns, the matrix A is of order n × n, also known as a square matrix (of order n).

Square matrices A have the property that in many cases an inverse A^{-1} can be found. When this is the case, then the inverse matrix can be applied to both sides of the matrix equation, yielding the solution for X. In other words, multiplying both sides by A^{-1} yields $A^{-1}AX = X = A^{-1}B$, **[See "eresources: Equations and Problems" on the Website mentioned at the beginning of the chapter.]**

In other words, the solution to the system of equations will result from finding A−1. The exact method for finding this can be found in any book on Linear Algebra, or also in our companion book "Behavioral Cybersecurity", Patterson and Winston-Proctor, CRC Press, 2019.

PROBLEMS

1. Solve the equations (or indicate if there is no solution):

   $x^3 = 2$ (mod 15)   Solution(s): _____

   $x^2 + x + 1 = 0$. (mod 17)   Solution(s): _____

   2^{10} (mod 18)   Solution(s): _____

   3^{1001} (mod 40)   Solution(s): _____

2. Solve $\sqrt{x} = 17$ (mod 29).

   Solution(s): _____

3. Consider $\mathbf{Z}_{21}$ modular arithmetic, or mod 21 arithmetic. List all of the possibilities for an equation $a \times b \equiv 0$ (mod 21), where neither a nor b is 0 itself.

   a. Create the multiplication table for $\mathbf{Z}_{15}$.

   b. Solve the following for $\mathbf{Z}_{15}$:

7×8	_____
4^{-1}	_____
$6^{-1} + (4 \times 7^{-1})$	_____
$8^2 \times 4^{-1}$	_____
$3 \div 8$	_____
$6 \div 5$	_____

4. Square elements in mod systems are interesting. Many non-zero elements in mod systems are not squares. Take as an example $\mathbf{Z}_7$. Note that 1, 2, and 4 are squares, because, for example, $6^2 = 1$ (mod 7), $3^2 = 2$ (mod 7), and $5^2 = 4$ (mod 7). Also, you can show that 3, 5, and 6 are not squares.

5. Find all of the (nonzero) squares and nonsquares in mod 12 (or $\mathbf{Z}_{12}$).

   Squares _____

   Nonsquares _____

6. Display the calculations to find

   GCD(8624,1837) = _____ and 1837^{-1} (mod 8624) = _____
   Or check if it doesn't exist _____

   GCD(89379,21577) = _____ and 21577^{-1} (mod 89379) = _____
   Or check if it doesn't exist _____

   GCD(438538,218655) = _____ and 218655^{-1} (mod 438538) = _____ Or check if it doesn't exist _____

7. Compute

   a. $3 \div 4$ (mod 17) Solution: _____ Or does not exist _____

   b. $18 \div 33$ (mod 121) Solution: _____ Or does not exist _____

   c. $27 \div 16$ (mod 43) Solution: _____ Or does not exist _____

   d. $12 \div 7$ (mod 15) Solution: _____ Or does not exist _____

8. Calculate the Möbius function for n = 77. Find all the elements in $\mathbf{Z}_{77}$ that are not relatively prime to n.

9. For each of x = 33, 46, 49, 67, find the smallest exponent a for which $x^a \equiv 1$ (mod n).

10. Multiply the following matrices (if this is possible):

2	3	1	5
−2	4	6	0

1	0	5
2	7	1
6	6	2
3	8	3

13

MODERN CRYPTOGRAPHY

Please see edition equations online at https://www.routledge.com/Behavioral-Cybersecurity-Fundamental-Principles-and-Applications-of-Personality/Patterson-Winston-Proctor/p/book/9780367509798.

In cryptography, the development of cryptographic techniques is inspired by what is called Kerckhoff's principle. This concept holds that almost all information about the cryptographic method should be revealed to the public, except for one component—the so-called *key* to the encryption. This long-established and reliable principle—also reestablished for the modern electronic era by Claude Shannon—is based on the concept that if we simply *hid* the cryptographic method, as soon as it was breached or obtained by bribery or other such means, the system would be compromised.

MODERN CRYPTOGRAPHIC TECHNIQUES

Once the digital age came upon us, the approach to cryptography changed dramatically. The understanding, once the communication of information became based on electronic transmission rather than on the printed word, was that what is referred to as the underlying alphabet of the system for encrypting and decrypting messages changed from the letters of a natural language (e.g., for the English language, {A, B, C, ..., X, Y, Z}) to the language of the digital era{0, 1}, the binary alphabet).

To underscore one difference between the two approaches, the structure of most human languages tends to differentiate in the usage of the underlying letters of such alphabets. For example, virtually any sufficiently large sample of text in English will demonstrate that the letter E will occur most often in the text, usually about 50% more often than the letter, usually T, that is the second most frequent (Patterson, 1987).

In our modern era, we will focus on two cryptosystems in some detail. A third, which dominated for many years, called the Data Encryption Standard (DES), is still in wide use but is no longer accepted as a standard.

The US government in the early 1970s came to realize that there was a need for an official US government standard for data encryption for civilian purposes.

A collaboration developed between the National Security Agency, National Bureau of Standards, and IBM to develop a proposal for a standard for data encryption, and it was published in 1975 as the DES. It used principles of both transposition and substitution, and it acted on a plaintext message divided into 64-bit pieces, with an encryption key of 56 bits and 16 individual transformations of successive 64-bit pieces to result in encrypted 64-bit messages (NBS, 1977).

From the publication of the DES in 1975 until the late 1990s, the DES was the only standard approved by the US government, but it was also the source of vast and heated differences in the crypto-community. In the 1990s, the US government realized that DES was substantially broken, so efforts were made to develop Advanced Encryption Standard (AES).

THE ADVANCED ENCRYPTION STANDARD

In 1997, the National Institute for Standards and Technology (NIST) began a process to establish a new standard. In a drastic reversal from the approach used in the development of the DES, a public call was issued throughout the world for the development of a new standard and inviting the submission of algorithms subject to some initial design criteria. Fifteen were selected for further review. It is interesting to note that 10 of the original 15 were not from the United States but from nine different countries.

The final selection, after considerable international review by the cryptographic community, was Rijndael (pronounced rain-doll), which was a submission by two Belgians, Vincent Rijmen and Joan Daemen (Rijndael is a fusion of their last names).

Upon its adoption—shortly after the "9/11" attack, it was renamed the Advanced Encryption Standard, accepted as a US government standard. The best reference is *The Design of Rijndael* (Daemen & Rijmen, 2001). The official government AES is found at the NIST (2001).

AES is based on a design principle known as a substitution–permutation network and is efficient in both software and hardware. AES is a variant of Rijndael that has a fixed block size of 128 bits and a key size of 128, 192, or 256 bits.

AES operates on a 4×4 column-major order array of bytes, termed the *state*. Most AES calculations are done in a particular finite field.

The key size used for an AES cipher specifies the number of transformation rounds that convert the input, called the plaintext, into the final output, called the ciphertext. The number of rounds is as follows:

- 10 rounds for 128-bit keys
- 12 rounds for 192-bit keys
- 14 rounds for 256-bit keys

In Rijndael, all of the mathematics can be done in a system called a Galois field. However, it is not essential to do computations in the Galois field, but simply to use the results by looking up into appropriate tables.

We begin with the initial text to be encrypted, broken into an appropriate number of bytes, in our case 16 bytes. This text, and each time it is transformed, will be called the State. We will normally represent the bytes throughout the algorithm as hexadecimal symbols, {0, 1, 2, 3, 4, 5, 6, 7, 8, 9, a, b, c, d, e, f}, where the six letter representations correspond to the decimal numbers 10, ..., 15. A hex or hexadecimal number (base 16) is represented by 4 bits—corresponding to the above {0000, 0001, 0010, 0011, 0100, 0101, 0110, 0111, 1000, 1001, 1010, 1011, 1100, 1101, 1110, 1111}. Since a byte is 8 bits, any byte can be represented by the 8 bits, and each half of the byte by 4 bits, or by two hexadecimal numbers. For example, the 8-bit representation of a certain byte may be 01011011, or 0101 1011 (two hex numbers), or 5b (hex). Also, 5b (hex) expressed as an integer is $(5 \times 16) + b = (5 \times 16) + 11 = 91$ (decimal).

A critical step in dealing with these hex numbers is to determine their logical XOR ($\oplus$) bit by bit or byte by byte. **[See "eresources: Equations and Problems" on the Website mentioned at the beginning of the chapter.]**

We include here a complete table of hex digits with the XOR or $\oplus$ operation.

Hex Table under XOR or ⊕ : Tables to Perform a Rijndael/ AES Encryption

⊕	0	1	2	3	4	5	6	7	8	9	A	B	C	D	E	F
0	0	1	2	3	4	5	6	7	8	9	A	B	C	D	E	F
1	1	0	3	2	5	4	7	6	9	8	B	A	D	C	F	E
2	2	3	0	1	6	7	4	5	A	B	8	9	E	F	C	D
3	3	2	1	0	7	6	5	4	B	A	9	8	F	E	D	C
4	4	5	6	7	0	1	2	3	C	D	E	F	8	9	A	B
5	5	4	7	6	1	0	3	2	D	C	F	E	9	8	B	A
6	6	7	4	5	2	3	0	1	E	F	C	D	A	B	8	9
7	7	6	5	4	3	2	1	0	F	E	D	C	B	A	9	8
8	8	9	A	B	C	D	E	F	0	1	2	3	4	5	6	7
9	9	8	B	A	D	C	F	E	1	0	3	2	5	4	7	6
A	A	B	8	9	E	F	C	D	2	3	0	1	6	7	4	5
B	B	A	9	8	F	E	D	C	3	2	1	0	7	6	5	4
C	C	D	E	F	8	9	A	B	4	5	6	7	0	1	2	3
D	D	C	F	E	9	8	B	A	5	4	7	6	1	0	3	2
E	E	F	C	D	A	B	8	9	6	7	4	5	2	3	0	1
F	F	E	D	C	B	A	9	8	7	6	5	4	3	2	1	0

The encryption key is also pictured as a rectangular array with four rows and columns of bytes.

In the case of our example, we will choose a text or State of 128 bytes, a key of similar size, and ten rounds in the encryption. The pseudo-C code for a round will be:

```
Round(State, ExpandedKey[i])
{
SubBytes(State);
ShiftRows(State);
MixColumns(State);
AddRoundKey(State, ExpandedKey[i]);
}
```

Let us take these four steps in order.

SubBytes

SubBytes is a simple byte-by-byte substitution from two tables that can be found at the AES (pp. 16 and 22).

Just for an example, suppose you wish to find the result of SubBytes for the byte [1010 0111] = a7. Look in the S_{RD} table (S-box) to find $S_{RD}(a7) = 5c = [0101\ 1100]$. Should you need to invert that result, look in the S_{RD}^{-1} table (inverse S-box, p. 23) to find $S_{RD}^{-1}(5c) = a7$.

ShiftRow

ShiftRow takes each row of the State and does a circular shift by 0, 1, 2, 3 positions as follows:

As opposed to SubBytes being a substitution, ShiftRow is a transposition. None of the byte values are changed, just the position of many of the bytes.

a	b	c	d		a	b	c	d
e	f	g	h	→	f	g	h	e
i	j	j	k	→	j	k	i	j
l	m	n	o	→	o	l	m	n

MixColumns

MixColumns introduces the main complexity in the overall algorithm. Viewed from the perspective of the underlying Galois fields, it is essentially a matrix multiplication. But without going into the algebra of Galois fields, we can carry out this computation also as a form of matrix multiplication but where the "multiplication" of individual elements is essentially a logarithmic and antilogarithmic substitution, called "mul," and the "addition" along a row and column is the bitwise or bytewise exclusive-or operation.

AddRoundKey

The final step in an AES round is the selection of the portion of the key for the next round.

We are using AES in the mode with 128-bit (16-byte) key and test messages and using the 10-round version.

Suppose the message or plaintext is as follows, in hex bytes:

```
32    43    f6    a8    88    5a    30    8d    31    31
98    a2    e0    37    07    34
```

and the key is:

```
2b    7e    15    16    28    ae    d2    a6    ab    f7
15    88    09    cf    4f    3c
```

The standard way of describing a Rijndael/AES encryption and decryption is with the terminology for each step being described as R[xx].yyyy, where the xx denotes a round, going from 00 to 10, and the yyyy represents the step, derived from the pseudocode:

R[00].input	Only for round 00. This is the plaintext to be encrypted
R[00].k_sch	For the particular round (00 to 10), this is the key being used for this round
R[01].start	Simply the XOR of the plaintext and key
R[01].s_box	The procedure called ByteSub using the single S-box
R[01].s_row	The result of a procedure called ShiftRow
R[01].m_col	The result of a procedure called MixColumn
R[01].k_sch	The Key Schedule, or the generated key for the next round
R[02].start	Again, the XOR of the result of round one and the key schedule generated at the end of round one

Pseudo-C code for a round:
Round(State, ExpandedKey[i])

Test Vectors

This example (found online) is chosen from Daemen and Rijmen (2001, pp. 215–216). For this example, we only compute one round.

This example assumes a 128-bit (or 16-byte) test message and cipher key.

Message or plaintext is (in hex bytes):
32 43 f6 a8 88 5a 30 8d 31 31 98 a2 e0 37 07 34
The key is:
2b 7e 15 16 28 ae d2 a6 ab f7 15 88 09 cf 4f 3c
Using the standard format for a trace of the encryption, we have:

```
R[00].input =
32 43 f6 a8 88 5a 30 8d 31 31 98 a2 e0 37 07 34
R[00].k_sch =
2b 7e 15 16 28 ae d2 a6 ab f7 15 88 09 cf 4f 3c
```

Now compute R[01].start by computing the XOR of R[00].input with R[00].k_sch:

```
32 43 f6 a8 88 5a 30 8d 31 31 98 a2 e0 37 07 34 ⊕
2b 7e 15 16 28 ae d2 a6 ab f7 15 88 09 cf 4f 3c =
19 3d e3 be a0 f4 e2 2b 9a c6 8d 2a e9 f8 48 08
```

So:

```
R[00].input =
32 43 f6 a8 88 5a 30 8d 31 31 98 a2 e0 37 07 34
R[00].k_sch =
2b 7e 15 16 28 ae d2 a6 ab f7 15 88 09 cf 4f 3c
R[01].start =
19 3d e3 be a0 f4 e2 2b 9a c6 8d 2a e9 f8 48 08
```

Computing R[01].s_box

This is the SubBytes or S-box step

Note the S-box table, S_{RD}, can be found at the end of the chapter, as well as its inverse, S_{RD}^{-1} (Table 13.1).

TABLE 13.1

SubByte S_{RD} and S_{RD}^{-1} Tables

S_{RD}									y							
		0	1	2	3	4	5	6	7	8	9	a	b	c	d	e	f
	0	52	09	6a	d5	30	36	a5	38	bf	40	a3	9e	81	f3	d7	fb
	1	7c	e3	39	82	9b	2f	ff	87	34	8e	43	44	c4	de	e9	cb
	2	54	7b	94	32	a6	c2	23	3d	ee	4c	95	0b	42	fa	c3	4e
	3	08	2e	a1	66	28	d9	24	b2	76	5b	a2	49	6d	8b	d1	25
	4	72	f8	f6	64	86	68	98	16	d4	a4	5c	cc	5d	65	b6	92
	5	6c	70	48	50	fd	ed	b9	da	5e	15	46	57	a7	8d	9d	84
	6	90	d8	ab	00	8c	bc	d3	0a	f7	e4	58	05	b8	b3	45	06
	7	d0	2c	1e	8f	ca	3f	0f	02	c1	af	bd	03	01	13	8a	6b
x	8	3a	91	11	41	4f	67	dc	ea	97	f2	cf	ce	f0	b4	e6	73
	9	96	ac	74	22	e7	ad	35	85	e2	f9	37	e8	1c	75	df	6e
	a	47	f1	1a	71	1d	29	c5	89	6f	b7	62	0e	aa	18	be	1b
	b	fc	56	3e	4b	c6	d2	79	20	9a	db	c0	fe	78	cd	5a	f4
	c	1f	dd	a8	33	88	07	c7	31	b1	12	10	59	27	80	ec	5f
	d	60	51	7f	a9	19	b5	4a	0d	2d	e5	7a	9f	93	c9	9c	ef
	e	a0	e0	3b	4d	ae	2a	f5	b0	c8	eb	bb	3c	83	53	99	61
	f	17	2b	04	7e	ba	77	d6	26	e1	69	14	63	55	21	0c	7d

Note: x and y represent the left and right hexadecimal numbers to be looked up. For example, for byte 9e, x = 9, y = e; and $S_{RD}(9e) = df$. Similarly, for byte 4a, x = 4, y = a; and $S_{RD}^{-1}(4a) = d6$.

(Continued)

TABLE 13.1 (Continued)
SubByte S_{RD} and S_{RD}^{-1} Tables

S_{RD}^{-1}		0	1	2	3	4	5	6	7	8	9	a	b	c	d	e	f
	0	63	7c	77	7b	f2	6b	6f	c5	30	01	67	2b	fe	d7	ab	76
	1	ca	82	c9	7d	fa	59	47	f0	ad	d4	a2	af	9c	a4	72	c0
	2	b7	fd	93	26	36	3f	f7	cc	34	a5	e5	f1	71	d8	31	15
	3	04	c7	23	c3	18	96	05	9a	07	12	80	e2	eb	27	b2	75
	4	09	83	2c	1a	1b	6e	5a	a0	52	3b	d6	b3	29	e3	2f	84
	5	53	d1	00	ed	20	fc	b1	5b	6a	cb	be	39	4a	4c	58	cf
	6	d0	ef	aa	fb	43	4d	33	85	45	f9	02	7f	50	3c	9f	a8
	7	51	a3	40	8f	92	9d	38	f5	bc	b6	da	21	10	ff	f3	d2
x	8	cd	0c	13	ec	5f	97	44	17	c4	a7	7e	3d	64	5d	19	73
	9	60	81	4f	dc	22	2a	90	88	46	ee	b8	14	de	5e	0b	db
	a	e0	32	3a	0a	49	06	24	5c	c2	d3	ac	62	91	95	e4	79
	b	e7	c8	37	6d	8d	d5	4e	a9	6c	56	f4	ea	65	7a	ae	08
	c	ba	78	25	2e	1c	a6	b4	c6	e8	dd	74	1f	4b	bd	8b	8a
	d	70	3e	b5	66	48	03	f6	0e	61	35	57	b9	86	c1	1d	9e
	e	e1	f8	98	11	69	d9	8e	94	9b	1e	87	e9	ce	55	28	df
	f	8c	a1	89	0d	bf	e6	42	68	41	99	2d	0f	b0	54	bb	16

Note: These two tables take a decimal input from 0 to 255, left to right. Express the input as $16a + b$; then the appropriate value is in row a and column b.

Logtable and Alogtable

LogTable

1	0	1	2	3	4	5	6	7	8	9	a	b	c	d	e	f
0	0	0	25	1	50	2	26	198	75	199	27	104	51	238	223	3
1	100	4	224	14	52	141	129	239	76	113	8	200	248	105	28	193
2	125	194	29	181	249	185	39	106	77	228	166	114	154	201	9	120
3	101	47	138	5	33	15	225	36	18	240	130	69	53	147	218	142
4	150	143	219	189	54	208	206	148	19	92	210	241	64	70	131	56
5	102	221	253	48	191	6	139	98	179	37	226	152	34	136	145	16
6	126	110	72	195	163	182	30	66	58	107	40	84	250	133	61	186
7	43	121	10	21	155	159	94	202	78	212	172	229	243	115	167	87
8	175	88	168	80	244	234	214	116	79	174	233	213	231	230	173	232
9	44	215	117	122	235	22	11	245	89	203	95	176	156	169	81	160

(Continued)

a	127	12	246	111	23	196	73	236	216	67	31	45	164	118	123	183
b	204	187	62	90	251	96	177	134	59	82	161	108	170	85	41	157
c	151	178	135	144	97	190	220	252	188	149	207	205	55	63	91	209
d	83	57	132	60	65	162	109	71	20	42	158	93	86	242	211	171
e	68	17	146	217	35	32	46	137	180	124	184	38	119	153	227	165
f	103	74	237	222	197	49	254	24	13	99	140	128	192	247	112	7

Alogtable

	0	1	2	3	4	5	6	7	8	9	a	b	c	d	e	f
0	1	3	5	15	17	51	85	255	26	46	114	150	161	248	19	53
1	95	225	56	72	216	115	149	164	247	2	6	10	30	34	102	170
2	229	52	92	228	55	89	235	38	106	190	217	112	144	171	230	49
3	83	245	4	12	20	60	68	204	79	209	104	184	211	110	178	205
4	76	212	103	169	224	59	77	215	98	166	241	8	24	40	120	136
5	131	158	185	208	107	189	220	127	129	152	179	206	73	219	118	154
6	181	196	87	249	16	48	80	240	11	29	39	105	187	214	97	163
7	254	25	43	125	135	146	173	236	47	113	147	174	233	32	96	160
8	251	22	58	78	210	109	183	194	93	231	50	86	250	21	63	65
9	195	94	226	61	71	201	64	192	91	237	44	116	156	191	218	117
a	159	186	213	100	172	239	42	126	130	157	188	223	122	142	137	128
b	155	182	193	88	232	35	101	175	234	37	111	177	200	67	197	84
c	252	31	33	99	165	244	7	9	27	45	119	153	176	203	70	202
d	69	207	74	222	121	139	134	145	168	227	62	66	198	81	243	14
e	18	54	90	238	41	123	141	140	143	138	133	148	167	242	13	23
f	57	75	221	124	132	151	162	253	28	36	108	180	199	82	246	1

Note: These two tables take a decimal input from 0 to 255, left to right. Express the input as 16a+b, then the appropriate value is in row a and column b.

The operation is simply to look up the S-box value for each byte in R[01].start. For example, the first hex pair of R[01].start is 19. The S-box, the element at row 1, column 9 is d4.

Thus, the R[01].s_box is:

d4 27 11 ae e0 bf 98 f1 b8 b4 5d e5 1e 41 52 30

So now we have:

```
R[00].input =
32 43 f6 a8 88 5a 30 8d 31 31 98 a2 e0 37 07 34
R[00].k_sch =
2b 7e 15 16 28 ae d2 a6 ab f7 15 88 09 cf 4f 3c
R[01].start =
19 3d e3 be a0 f4 e2 2b 9a c6 8d 2a e9 f8 48 08
```

```
R[01].s_box =
d4 27 11 ae e0 bf 98 f1 b8 b4 5d e5 1e 41 52 30
```

Computing R[01].s_row:

This is the ShiftRows step. One writes R[01].s_box into a 4-by-4 array, column-wise (i.e., fill the first column first, then the second column, ...)

Writing Column-Wise				Now a Circular Left Shift			
d4	e0	b8	1e	d4	e0	b8	1e
27	bf	b4	41	bf	b4	41	27
11	98	5d	52	5d	52	11	98
ae	f1	e5	30	30	ae	f1	e5

Shift left by x positions in row x (x = 0, 1, 2, 3). Now write this out in a single row to get R[01].s_row:

d4 bf 5d 30 e0 b4 52 ae b8 41 11 f1 1e 27 98 e5

In the previous notation,

```
R[00].input =
32 43 f6 a8 88 5a 30 8d 31 31 98 a2 e0 37 07 34
R[00].k_sch =
2b 7e 15 16 28 ae d2 a6 ab f7 15 88 09 cf 4f 3c
R[01].start =
19 3d e3 be a0 f4 e2 2b 9a c6 8d 2a e9 f8 48 08
R[01].s_box =
d4 27 11 ae e0 bf 98 f1 b8 b4 5d e5 1e 41 52 30
R[01].s_row =
d4 bf 5d 30 e0 b4 52 ae b8 41 11 f1 1e 27 98 e5
```

Computing R[01]m_col

This is the MixColumns step, undoubtedly the trickiest:

- This is actually a computation in the Galois field of polynomials over GF(2,8), that is, polynomials of degree 7, with binary coefficients.
- But let us not worry about that. It can also be expressed as a matrix product of a fixed matrix C. Suppose the State after ShiftRow is R[01].s_row written column-wise:

C				R[01].s_row			
02	03	01	01	d4	e0	b8	1e
01	02	03	01	bf	b4	41	27
01	01	02	03	5d	52	11	98
03	01	01	02	30	ae	f1	e5

The result of this multiplication will be, as you know, another 4×4 matrix. Again, when we string out the column-wise version, we will get R[01].m_col.

However, these multiplications are in GF(2,8), or mod 256 arithmetic, and one can generate the "log tables" (see Rijmen & Daemen, 2001, pp. 221–222) to make the computation simpler. The relevant Logtable and anti-logtable (Alogtable) can be found at the end of this chapter.

Indeed, in the code is a brief function to do the multiplication (mul, p. 223).

Essentially, mul is

Alogtable[(Logtable[a] + Logtable[b]) (mod 256)]

We will only compute the first byte of the matrix product, which is found by the usual method of the first row of the left-hand matrix by the first column of the right-hand matrix, thus:

02	03	01	01	d4	e0	b8	1e
01	02	03	01	bf	b4	41	27
01	01	02	03	5d	52	11	98
03	01	01	02	30	ae	f1	e5

Yielding for the first component:

02 d4 $\oplus$ 03 bf $\oplus$ 01 5d $\oplus$ 01 30 = 02 d4 $\oplus$ 03 bf $\oplus$ 5d $\oplus$ 30 (01 is the identity)

Using the mul function for the first two terms (the right-hand side will be decimal numbers); that is, d4 in decimal is $13 \times 16 + 4 = 212$, and bf in decimal is $11 \times 16 + 15 = 191$:

mul(2,d4) = Alogtable[Logtable[2] + Logtable[212]]

$\qquad$ = Alogtable[25 + 65] = Alogtable[90] = 179 = $b3_{hex}$;

mul(3,bf) = Alogtable[Logtable[3] + Logtable[191]]

$\qquad$ = Alogtable[1 + 157] = Alogtable[158] = 218 = da_{hex}.

Then, we need to compute b3 $\oplus$ da $\oplus$ 5d $\oplus$ 30:

Or,			
b	1011	3	0011
d	1101	a	1010
5	0101	d	1101
3	0011	0	0000
Or			
	0000 = 00		0100 = 04

After the full matrix multiplication, we have:

```
R[00].input =
32 43 f6 a8 88 5a 30 8d 31 31 98 a2 e0 37 07 34
R[00].k_sch =
2b 7e 15 16 28 ae d2 a6 ab f7 15 88 09 cf 4f 3c
R[01].start =
19 3d e3 be a0 f4 e2 2b 9a c6 8d 2a e9 f8 48 08
R[01].s_box =
d4 27 11 ae e0 bf 98 f1 b8 b4 5d e5 1e 41 52 30
R[01].s_row =
d4 bf 5d 30 e0 b4 52 ae b8 41 11 f1 1e 27 98 e5
R[01].m_col =
04 66 81 e5 e0 cb 19 9a 48 f8 d3 7a 28 06 26 4c
```

Last step—key schedule.

In the key schedule, we use the previous key, XOR with another part of the previous key, run through the S-box, with a possible counter added.

This time, we will only calculate the first 4 bytes of the key.

Take the first 4 bytes of the former key: 2b 7e 15 16.

Left rotate once the last 4 bytes: 09 cf 4f 3c → cf 4f 3c 09.

Run this last part through the S-box: SubByte(cf 4f 3c 09) = 8a 84 eb 01.

XOR these, with a counter of 1 on the first byte:

2b $\oplus$ 8a $\oplus$ 01   7e $\oplus$ 84   15 $\oplus$ cb   16 $\oplus$ 01 = a0 fa fe 17.

The rest of the key schedule:

- On the previous slide, we determined the first 4 bytes of R[01].k_sch, namely: a0 fa fe 17.
- The other 12 bytes are gotten by XORing 4 bytes at a time from the previous key, R[00].k_sch and the new key, as follows:

- R[00].k_sch = 2b 7e 15 16 28 ae d2 a6 ab f7 15 88 09 cf 4f 3c
- R[01].k_sch = a0 fa fe 17 88 54 2c b1 23 a3 39 39 2a 6c 76 05.

```
R[00].input =
32 43 f6 a8 88 5a 30 8d 31 31 98 a2 e0 37 07 34
R[00].k_sch =
2b 7e 15 16 28 ae d2 a6 ab f7 15 88 09 cf 4f 3c
R[01].start =
19 3d e3 be a0 f4 e2 2b 9a c6 8d 2a e9 f8 48 08
R[01].s_box =
d4 27 11 ae e0 bf 98 f1 b8 b4 5d e5 1e 41 52 30
R[01].s_row =
d4 bf 5d 30 e0 b4 52 ae b8 41 11 f1 1e 27 98 e5
R[01].m_col =
04 66 81 e5 e0 cb 19 9a 48 f8 d3 7a 28 06 26 4c
R[02].start =
a4 9c 7f f2 68 9f 35 2b 6b 5b ea 43 02 6a 50 49
```

Getting to R[02].start consists of XORing R[01].m_col and R[01].k_sch.

Continue the same process for rounds 2, 3, …, 10.

SYMMETRIC ENCRYPTION OR PUBLIC KEY CRYPTOLOGY

Despite the pluses or minuses of AES, or any private key or symmetric method, one problem AES can never solve is the key management problem. Suppose we have a network with six users, each one of whom must have a separate key to communicate with each of the other five users. Thus, we will need in all for the six users (Figure 13.1).

THE PKC MODEL FOR KEY MANAGEMENT

Now consider the case of the following approach. For each of the 1000 users, choose a key $k_{i} = (kp_i, ks_i)$, $i = 1, …, 1000$. In a system-wide public directory, list all of the "public" keys kp_i, $i = 1, …, 1000$. Then, to send a message m to user j, select the public key, kp_i, and apply the encryption transformation $c = T(kp_i, m)$.

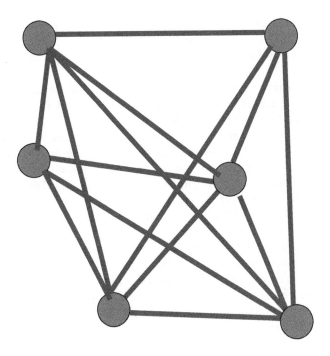

Figure 13.1 Complete graph with 6 nodes and 15 edges.

Send the ciphertext, c.

Only user j has the rest of the key necessary to compute the decryption:

$$T((kp_i, ks_i), c) = m$$

Thus, rather than having to manage the secret distribution of $O(n^2)$ keys in a network of n users, only n keys are required, and they need not be distributed secretly.

Furthermore, the public key concept could also be used for the authentication of messages in a way that a secret-key system could not address.

CAN WE DEVISE A PKC?

Therefore, if we could devise a PKC, it would certainly have most desirable features. But many questions remain to be asked.

First of all, can we devise a PKC? What should we look for? Second, if we can find one, will it be secure? Will it be efficient?

For now, we will consider only the general parameters of finding PKCs.

This approach, as described earlier, implies that the sender and receiver of encrypted information have two different parts of the key. Since the key k is broken into the so-called public key (k_p) and the secret key (k_s), and the public part is made available in the open for anyone to send a message to the creator of the key, and the secret key never leaves the creator, the creator and everyone else in the universe have different sets of information about the key, and therefore, such a method is referred to as "asymmetric encryption."

Unfortunately, defining the problem in this way, which will certainly solve the key management problem since only the public parts of everyone's key need be shared and can be placed in an open directory, and the private parts never have to leave the creator. But this is only a model and does not describe how this asymmetric approach can be effectively carried out. This problem was solved by three cryptographers, Ron Rivest, Adi Shamir (from MIT), and Len Adelman from the University of Southern California, and was called the RSA public key cryptosystem (Rivest et al., 1978).

THE RSA PUBLIC KEY CRYPTOSYSTEM

About 30 years ago, Rivest, Shamir, and Adleman developed their public key cryptosystem called RSA (wonder why?).

WHAT IS THE RSA CRYPTOSYSTEM?

It is simple enough that the RSA Security Company has put it on a t-shirt, whose text read:

$$\boxed{\begin{array}{c} \textbf{RSA Algorithm} \\ \textbf{P \& Q PRIME} \\ \textbf{N = PQ} \\ \textbf{ED} \equiv \textbf{1 MOD (P}-\textbf{1)(Q}-\textbf{1)} \\ \textbf{C} \equiv \textbf{M}^{\textbf{E}} \textbf{ MOD N} \\ \textbf{M} \equiv \textbf{C}^{\textbf{D}} \textbf{ MOD N} \end{array}}$$

- Take two prime numbers p and q (of 200 digits), and multiply n = pq.
- Find e and d such that their product gives a remainder of 1 when divided by (p − 1)(q − 1).
- To encrypt, raise the message m to the power e (mod n).
- To decrypt, raise the cipher c to the power d (mod n).

The details of the computations involved may be found in our companion book, *Behavioral Cybersecurity*, Patterson and Winston-Proctor, CRC Press, 2019.

PROBLEMS

1. Find the SubBytes transformations for the bytes a5, 3f, 76, c9. Verify your results using the inverse SubBytes table/

2. Show the AES MixColumns transformation for

C				R[01].s_row			
02	03	01	01	ef	7c	29	0b
01	02	03	01	c5	a6	33	d9
01	01	02	03	41	7a	99	e3
03	01	01	02	29	af	37	0d

3. Compute the indicated steps of an AES/Rijndael encryption when the message and key are given as follows:

```
R[00].input =
01 02 03 04 01 02 03 04 01 02 03 04 01 02 03 04
R[00].k_sch =
f0 f0 f0 f0 f0 f0 f0 f0 f0 f0 f0 f0 f0 f0 f0 f0
R[01].start =
__ __ __ __ __ __ __ __ __ __ __ __ __ __ __ __
R[01].s_box =
__ __ __ __ __ __ __ __ __ __ __ __ __ __ __ __
R[01].s_row =
__ __ __ __ __ __ __ __ __ __ __ __ __ __ __ __
R[01].m_col =
__ __ __ __ __ __ __ __ __ __ __ __ __ __ __ __
R[01].k_sch =
__ __ __ __ __ __ __ __ __ __ __ __ __ __ __ __
```

```
R[02].start =
```
— — — — — — — — — — — — — — — — — —

4. We know that an RSA cryptosystem can be broken if the prime numbers p and q are small enough.

   ```
   e = 325856364942268231677035294174763975263
   n = 661779642447352063488503270662016140733
   ```

5. Convert 167845 decimal to binary.
6. Use the fast exponentiation algorithm to compute x^{167845}. Show all the steps.

REFERENCES

Daemen, J. & Rijmen, V. 2001. *The Design of Rijndael.* Springer-Verlag, Berlin.

National Bureau of Standards (NBS). 1977. *Data Encryption Standard.* FIPS-Pub 46. NBS, Washington, DC.

National Institute for Standards and Technology (NIST). 2001. *Announcing the Advanced Encryption Standard (AES).* Federal Information Processing Standards Publication 197, November 26. https://csrc.nist.gov/csrc/media/publications/fips/197/final/documents/fips-197.pdf.

Patterson, W. 1987. *Mathematical Cryptology.* Rowman and Littlefield, Totowa, NJ.

Rivest, R., Shamir, A., & Adleman, L. 1978. A method for obtaining digital signatures and public-key cryptosystems (PDF). *Communications of the ACM*, 21(2), 120–126. doi: 10.1145/359340.359342.

14

STEGANOGRAPHY AND RELATION TO CRYPTO

As opposed to the definition of cryptography, *steganography* means *covered writing* (or originally in the Greek language, stegano, "στεγανο," which translates literally as "watertight" and, as in the definition of cryptography, the "-graphy" from γραφια—grafia or *writing*). Its history is as deep as is crypto. As a linguistic anomaly, the "watertight" translation is reminiscent of a usage of steganography in modern times, the creation of "watermarks" to identify a paper manufacturer.

There have been many methods over time of steganographs (or steganograms) that conceal the existence of a message. Among these are invisible inks, microdots, character arrangement (other than the cryptographic methods of permutation and substitution), digital signatures, covert channels, and spread-spectrum communications. As opposed to cryptography, steganography is the art of concealing the existence of information within innocuous carriers.

A message in ciphertext may arouse suspicion, while an *invisible* message will not. As a shorthand for the differences, cryptographic techniques "scramble" messages so if intercepted, the messages cannot be understood; steganography "camouflages" messages to hide their existence.

This one fact, in and of itself, suggests that the interface between cryptography and steganography needs to be explored within the context of behavioral science, since the approaches to creating and/or defending crypto or stego depend on human decisions based on their behavior.

A HISTORY OF STEGANOGRAPHY

One of the first documents describing steganography is from the *Histories* of Herodotus. In ancient Greece, text was written on wax-covered tablets. In one story, Demeratus wanted to notify Sparta that

Xerxes intended to invade Greece. To avoid capture, he scraped the wax off the tablets and wrote a message on the underlying wood. He then covered the tablets with wax again. The tablets appeared to be blank and unused so they passed inspection by sentries without question.

Another ingenious method was to shave the head of a messenger and tattoo a message or image on the messenger's head. After allowing his hair to grow, the message would be undetected until the head was shaved again. In modern parlance, this would be a pretty low-resolution methodology—perhaps a month to communicate a few bytes.

More common in more recent times, steganography has been implemented through the use of invisible inks. Such inks were used with much success as recently as World War II. Common sources for invisible inks are milk, vinegar, fruit juices, and urine, all of which darken when heated. These liquids all contain carbon compounds. When heated, the compounds break down and carbon is released, resulting in the chemical reaction between carbon and oxygen, that is, oxidation. The result of oxidation is a discoloration that permits the secret ink to become visible.

Null ciphers (unencrypted messages) were also used. The real message is "camouflaged" in an innocent-sounding message. Due to the "sound" of many open coded messages, the suspect communications were detected by mail filters. However "innocent" messages were allowed to flow through. An example of a message containing such a null cipher follows.

Suppose an obscure story appears on page 27 of the Sports section of the Oakland Tribune (CA):

> *However the baseball Athletics play ball, relievers cannot meet enviable needs for passing Houston's formidable array when winning under needy circumstances.*

An alert reader might uncover the steganogram conveying Winston Churchill's famous exhortation:

> *HoWever thE baSeballAtHletics plAy ball, reLievers caNnot meEt enViable neEds foR paSsing HoUston's foRmidable arRay whEn wiNning unDer neEdy ciRcumstances.*

That is, taking the third letter of each word:

> *We shall never surrender!*

The following message was actually sent from New York by a German spy in World War II:

> Apparently neutral's protest is thoroughly discounted and ignored. Isman hard hit. Blockade issue affects pretext for embargo on byproducts, ejecting suets and vegetable oils.

Taking the second letter in each word, the following message emerges:

> *APparently nEutral's pRotest iS tHoroughly dIscounted aNd iGnored. ISman hArd hIt. BLockade iSsue aFfects pRetext fOr eMbargo oN bYproducts, eJecting sUets aNd vEgetable oIls.*

Or,

> *Pershing sails from NY June 1.*

The Germans developed microdot technology, which FBI Director J. Edgar Hoover referred to as "the enemy's masterpiece of espionage" (Hoover, 1946).

TRANSMISSION ISSUES

Despite the long and interesting history of these various methods for hiding information, in practice they are declining in their importance.

One reason, although probably not the primary one, is the consequence of the use of a physical material to transmit information. Using invisible ink, to take one example, assumes that we have some medium on which this ink is deposited. Classically, this may be a letter with an innocuous message written on paper, and the invisible ink on top. Then the letter must reach its target, perhaps by postal service or courier. But who in these times would transmit such information on paper when electronic transmission is virtually instantaneous and capable of vastly larger messages or bandwidth?

Indeed, it may well be that the mere fact of transmission by a mail courier may arise suspicion, assuming the electronic means are readily available. Examples might include the threat a few years ago of anthrax contained in an envelope mailed to addresses in Washington, DC, including US senators, and also the more recent example of bombs sent through the mail to former presidents Obama and Clinton.

IMAGE STEGANOGRAPHY

Consequently, with the current electronic age, the field of steganography has shifted to techniques of converting the secret message into a bitstring and then injecting the bitstring bit by bit not into text, but into some other file format such as an image file (e.g., JPEG, TIFF, BMP, or GIF) or a sound or movie file (MPEG, WAV, or AVI).

There are usually two type of files used when embedding data into an image. The innocent-looking image that will hold the hidden information is a "container." A "message" is the information to be hidden. A message may be plaintext, ciphertext, other images, or anything that can be embedded in the least significant bits (LSBs) of an image.

In this environment, for example, in an image file, the altering of a single bit in the image may be impossible to detect, certainly to the human eye, but also to an analysis of the file content byte by byte.

IMAGE FILE FORMATS

The size of image files correlates positively with the number of pixels in the image and the color depth (bits per pixel). Images can be compressed in various ways, however. As indicated earlier, a compression algorithm stores either an exact representation or an approximation of the original image in a smaller number of bytes that can be expanded back to its uncompressed form with a corresponding decompression algorithm. Images with the same number of pixels and color depth can have very different compressed file sizes.

For example, a 640-by-480 pixel image with 24-bit color would occupy almost a megabyte of space:

$$640 \times 480 \times 24 \ = \ 7,372,800 \text{ bits} = 921,600 \text{ bytes} = 900 \text{ kB}$$

The most common image file formats are as follows:

JPEG (Joint Photographic Experts Group) is a lossy compression method. Nearly every digital camera can save images in the JPEG format, which supports 8-bit grayscale images and 24-bit RGB color images (8 bits each for red, green, and blue). JPEG lossy compression can result in a significant reduction of the file size. When not too great, the compression does not noticeably affect or detract from the image's quality, but JPEG files suffer generational degradation when repeatedly edited and saved.

The same image, when saved in the four image formats indicated earlier, requires considerably different space in kilobytes:

JPEG	712 kB
TIFF	6597 kB
BMP	18,433 kB
GIF	2868 kB

It is estimated that the human eye can distinguish perhaps as many as 3 million colors. If we use a fairly common color scheme or palette for an image to hide the steganograph, namely JPEG, there are 16.77 million possible RGB color combinations. Thus, we have many choices for altering the byte value associated with a pixel in order to conceal many bits of information and yet leave the image indistinguishable to the human eye.

Of course, if you have both the original image and the altered image, you do not have to rely on the human eye. You can use a "hex editor" (Hörz, 2018) to examine both the original image and the altered image byte by byte, and then it is a simple task to detect the differences.

It should also be noted: a JPEG file of a 4×6 image might be on the order of megabyte. A text to insert might be several kilobytes.

AN EXAMPLE

Example: We will use the easily available software tools HxD (Hörz, 2018) and QuickStego (Cybernescence, 2017).

We can present an example using two software tools, which any reader can download by himself or herself. These are both freeware products. One is a hexadecimal or hex editor, which allows the user to examine any file—therefore an image file byte by byte. There are many such editors available. One that we have used for this example is called HxD. The other software necessary for this example is again freeware, to insert a text into an image (thus creating a steganogram) where there are also numerous examples. The one we have chosen is called QuickStego.

It is often the case to use 256-color (or grayscale) images. These are the most common images found on the Internet in the form of GIF files. Each pixel is represented as a byte (8 bits). Many authors of steganography software stress the use of grayscale images (those with 256 shades of gray or better). The importance is not whether the image is grayscale; the importance is the degree to which the colors change between bit values.

Grayscale images are very good because the shades gradually change from byte to byte.

USING CRYPTOGRAPHY AND STEGANOGRAPHY IN TANDEM OR IN SEQUENCE

As we have seen, the cryptographic approach to secure messaging and the steganographic approach operate under two distinctly different and contradictory approaches. Cryptography is very open in telling any opponent even the technique or algorithm that is being used. In fact, in the RSA approach to public-key cryptology, any attacker can readily determine the size of the challenge in breaking the code because of the partial information involved in the public key. On the other hand, the steganographic approach attempts to appear completely normal, in the hopes that the attacker will be led to believe that there is no secret messaging involved and therefore will decide not to employ methods to try to determine if there is some secret to be revealed.

Even though these two approaches would seem to imply that the user must choose one or the other, there is a theory emerging that the two approaches of crypto and stego might be used in combination in various fashions.

Here is a very simple example. Suppose a person wishes to warn an ally of an impending attack. This person creates an image that clearly has the message "ATTACK AT DAWN." Then this person will send this to an ally. However, the message that will be clearly seen by anyone intercepting a message will warn of such an impending attack. However, what the sender has actually done, using, for example, the software State of, is embed the message "ATTACK AT MIDNIGHT."

The example is described in Figure 14.1.

Another more extensive example was presented in a recent master's thesis (Kittab, 2016) where the author, calling his approach Matryoshka steganography, used five levels of embedding in the

Figure 14.1 Hiding a stego message "ATTACK AT MIDNIGHT" in a graphic image and viewing part of the clear and the stego in a hex editor.

hopes of deterring the person intercepting the message from burrowing five levels deep in order to find the true message. This is a relatively new area of research and points out a more important aspect of how human behavior factors into all areas of cybersecurity.

The human factor involved in the use of cryptology arises when the defender, while telling the attacker everything but the key, relies upon the attacker's state of mind: Since we assume that the attacker has as great a knowledge of cryptology as the defender, the attacker can make a very determined calculation as to the cost of launching an attack—that is, trying to break the encryption—or to let it go by as not being worth the cost involved in deploying resources trying to break it.

Then, together with a factor based on a calculated risk, the choice of security mechanism might be balanced between the crypto strategy, the stego strategy, or indeed a combination of both.

COMMENTS

Steganography has its place in security. It is not intended to replace cryptography but supplement it. As we will see, the existence of both cryptography and steganography leads to the possibility of hybrid techniques, which we will examine more deeply in the next chapter. Hiding a message with steganography methods reduces the chance of a message being detected. If that message is also encrypted, if discovered, it must also be cracked (yet another layer of protection). There are very many steganography applications. Steganography goes well beyond simply embedding text in an image. It does not only pertain to digital images but also to other media (files such as voice, other text and binaries; other media such as communication channels, and so on).

PROBLEMS

1. Create a steganogram that embeds the following 20-letter message in a larger text by using the same letter position in each word, either the first, second, third, or fourth. For example, if you were creating a steganogram for the message "HELLO," it might be "Have Everyone Leave Liquor Outside." Once you choose the letter position, all the words in your steganogram must use the same position, as in the example. (Ignore the blanks in the message in the following.)
   FOURSCORE AND SEVEN YEARS AGO
2. Consider the principle of lossless versus lossy compression. Suppose you have an image wherein about 80% of the pixels represent the same color. How could you develop a coding system so that you would preserve all of the locations of the bytes of the same color, yet save a good deal of space in the rendering of the image?
3. Comment on the options available for Angus and Barbara in trying to establish a secure mechanism to exchange steganographs.
4. Create your own ATTACK AT DAWN, ATTACK AT MIDNIGHT example using QuickStego. Using a hex editor such as HxD, find all byte differences.

5. Manually embed a 30-byte message in a JPEG.
6. Use HxD to find all byte differences from your example in problem 2, from the position of the first difference

REFERENCES

Cybernescence. 2017. QuickStego. http://www.quickcrypto.com/free-steganography-software.html.

Hoover, J. E. 1946. The enemy's masterpiece of espionage. *Reader's Digest*, 48, 1–6.

Hörz, M. 2018. HxD—Freeware Hex Editor and Disk Editor. https://mh-nexus.de/en/hxd/.

Kittab, W. M. 2016. Matryoshka Steganography. M.Sc. thesis, Howard University.

15

A METRIC TO ASSESS CYBERATTACKS

Many individuals are now confused about what measures they can or should take in order to protect their information. We can give a model—in a very limited circumstance—where that level of protection can be determined very precisely. It should be noted that the research developed in this chapter has had the invaluable assistance of one of the leading cybersecurity researchers in Africa, Professor Michael Ekonde Sone of the University of Buea in Cameroon (Sone & Patterson, 2017).

DEFINING A CYBERSECURITY METRIC

One of the main problems in trying to develop a definitive metric to describe a cybersecurity scenario that takes into account human decision-making is the difficulty in trying to measure the cost to either an attacker or defender of a specific approach to a cybersecurity event.

We can, however, describe one situation in which it is possible to describe very precisely the cost of engaging in a cybersecurity attack or defense. This example is somewhat unique in terms of the precision or cost/benefit.

The case in point will be the use of the RSA public key cryptosystem.

THE ATTACKER/DEFENDER SCENARIO

In terms of the overall approach to the problem, it is necessary to define a metric useful to both parties in a cyberattack environment, whom we will call the attacker and the defender. An attacker may be an individual, an automated "bot," or a team of intruders.

The classic challenge we will investigate will be formulated in this way: The attacker is capable of intercepting various communications from the defender to some other party, but the intercepted message is of little use because it has been encrypted.

With any such encrypted communication, which we also call ciphertext, we can define the cost of an attack (to the attacker) in many forms. Assuming the attack takes place in an electronic environment, then the cost to decipher the communication will involve required use of CPU time to decipher, or the amount of memory storage for the decryption, or the attacker's human time in carrying out the attack. Each of these costs can essentially be exchanged for one of the others. Let us assume that we will standardize the cost of an attack or defense by the computer time necessary for the attack.

RIVEST–SHAMIR–ADLEMAN: AN INTERESTING EXAMPLE

There is one example where, despite many efforts over the past 40 years, it is possible to determine precisely the cost to an attacker to break a given encryption. This is the encryption method known as the RSA (see Chapter 13.9).

The ingenuity in the definition of the RSA is that a very simple algebraic process can lead to the only known method of breaking the encryption through the factoring of a number n that is the product of two primes, p and q. It is also the case that there is a wide latitude in choosing a number n that may have any number of decimal digits. For numbers n that will provide sufficient security, we will allow n to have between 70 and 200 decimal digits, i.e., $10^{70} < n < 10^{200}$.

Fortunately for this discussion, the RSA cryptosystem is highly scalable in that it can begin with an arbitrary number of digits for the choice of primes p and q (although p and q should be roughly of the same size). Consequently, the designer of a crypto defense can choose these parameters and therefore parameterize the cost to a potential attacker based on the number of digits of p and q. Furthermore, because the n = pq is known to all as part of the algorithm, the attacker also can determine the costs in launching an attack based on factoring the number n.

Fortunately, there is a great deal of literature on the cost of factoring such a number (Pomerance, 1996). To date, the best known factoring algorithms are of:

$$O\left(e^{\left((64b/9)^{1/3} (\log b)^{2/3} \right)} \right) \qquad (15.1)$$

where b is the number of bits of n expressed as a binary.

Let us consider the range of values for the public key, n, that can be productively used in an attack/defense scenario. First, since p and q, the only prime factors of n, should be approximately the same size, say M decimal digits, then $\log_{10} n \approx 2M$ decimal digits. So we can choose n so that $\log_{10} n$ can be any even integer.

Furthermore, in the complexity of the best known factoring method, the actual time will be some constant times $e^{\left((64b/9)^{1/3}(\log b)^{2/3}\right)}$, where b is the number of bits of n expressed as a binary.

The constant will be determined by external factors unique to the factoring attempt, such as multithreaded computing, speed of the processor, and so on. Nevertheless, relatively speaking, the overall cost of the factoring attempt will not vary greatly.

If we compute the predicted run times over a large range, using all n values from $10^{70} < n < 10^{350}$, we can predict within a constant the cost of an attack for any value of n in the range.

For large values of $\log_{10} n$, it is more descriptive of the attacker's cost when using a log plot. For example, a few values calculated are as follows: If the $\log_{10} n$ is, respectively, 70, 100, 130, or 200 decimal digits, the time to factor an integer of those magnitudes are 3373 seconds, 439.4 hours, 10.4 years, or approximately 247,000 years, respectively.

For a strong defense, the size of n will certainly need to be greater than 70. However, the function

$$O\left(e^{\left((64b/9)^{1/3}(\log b)^{2/3}\right)}\right)$$

grows in faster than polynomial time. So the runtime to factor n will be given by the following best fit linear function: $y = (3.01720 \times 10^{12})\,x - 2.26540 \times 10^{15}$, with a positive correlation coefficient of 0.95. So this linear fit will give a good (enough) first approximation to the time to factor for any n in a sufficiently large range.

ATTACK/DEFENSE SCENARIOS

Unfortunately, this example so far is almost unique in trying to establish a reliable metric in terms of potential attacks or defenses in the cybersecurity environment. The reason for this is that with many other security measures, there is not the same reliability on a specific nature of the attack that might be attempted, and thus the measures cannot be so reliably determined.

The simplicity of RSA encryption lends itself to the type of analysis described earlier and the type of metric that can be assigned to it because of a number of characteristics of the RSA. First, RSA is virtually completely scalable, so that a defender establishing an RSA system, as long it is done properly, can make the choice of almost any size integer n as the base of the public information. Second, the choice of n can be taken based on the computational cost of trying to factor that number n, a cost that is very well known and quite stable in terms of research on factoring. Next, despite approximately 40 years of effort in trying to find other ways of breaking RSA, factoring the number n remains the only known way of breaking the cryptosystem.

Thus, the example described in this chapter hopefully provides a methodology for establishing the strategy for a successful defense strategy based on the precision of the metric that has been designed in this chapter.

In essence, then, the use of RSA, under the control of the defender, can determine precisely the cost to an attacker of trying to break the encryption, and so it implies that the defender can thus issue the challenge, knowing that both attacker and defender will be able to calculate the cost of a successful attack. Let us consider several examples:

First scenario: The intent of the defender is simply to deter the attacker from any attempt to use a factoring algorithm, because the calculation of the cost will be so high that no rational attacker will bother to attempt the attack, knowing that the cost will be beyond any potential value of the resource to be stolen, or the attacker's ability to assume the cost for successful attack. Thus, the reasonable attacker will not even bother to try the attack.

The defender will assess the value of his or her information, and from the analysis, choose the appropriate value for n, say $n(D)$. This sets a lower bound for D's actual choice of n. Next, D will estimate A's ability to attack, then convert to a $n(A)$, and choose for the actual n the value so that $n > \max(n(A), n(D))$.

For this scenario, the defender will choose values of n in the range $160 \leq \log_{10} n \leq 200$. The cost involved will deter the attacker even if he or she is using a multipurpose attack with k processors.

Second scenario: The defender will deliberately choose the cost of the attack to be so low that any and every potential attacker will determine the cost to be acceptable and thus proceed to solve the factoring problem. This would not seem to be the best strategy for the defender; however, it may be that the defender has deliberately decided to allow the attacker to enter because upon retrieving the decrypted information, the attacker may be led into a trap. This, in cybersecurity technology, is often referred to as a honeypot.

In this case, regardless of the value of D's information, D will estimate A's ability to attack as $n(D)$ and then deliberately choose $n < n(D)$.

Third scenario: The defender chooses a difficulty level for the RSA that will cause the potential attacker to make a judgment call about whether the cost of an attack would be expected to be less than the projected value of the information obtained in a successful attack. In this way, the defender and attacker essentially establish a game theory problem, whereby both the attacker and defender will need to establish a cost and benefit of the information being guarded.

CONCLUSION

We can continue to look for other measures to establish the cost to attack or defend, but this example is so far almost unique. Despite 40 years of effort in trying to break RSA, factoring the number n remains the only known way of doing so.

PROBLEMS

1. Suggest an attack on the 20-letter alphabet encryption method described earlier that would be an improvement on trying all 20! keys.
2. In Chapter 5, we introduced the ABCD method of classifying hackers. Suppose you knew which type of hacker you had to defend against. Using the aforementioned terminology, what magnitude of prime products (n) would you choose if you were facing an A, B, C, or D attacker?
3. Comment on the scenarios described earlier.

REFERENCES

Patterson, W. 2016. Lecture Notes for CSCI 654: Cybersecurity I. Howard University, September, Washington, DC, USA.

Pomerance, C. 1996 . A tale of two sieves. *Notices of the AMS*, 43(12), 1473–1485.

Rivest, R., Shamir, A., & Adleman, L. 1978. A method for obtaining digital signatures and public-key cryptosystems (PDF). *Communications of the ACM*, 21(2), 120–126. doi: 10.1145/359340.359342.

Sone, M. E., & Patterson, W. 2017. *Wireless Data Communication: Security-Throughput Tradeoff.* Editions Universitaires Européennes, Saarbrücken.

16

BEHAVIORAL ECONOMICS

Behavioral economics is the study of how psychological, cognitive, emotional, cultural, and social factors impact the economic decisions of individuals and institutions and how these factors influence market prices, returns, and resource allocation. It also studies the impact of different kinds of behavior in environments of varying experimental values. An overview of the topic can be found on Wikipedia (2018).

It is primarily concerned with the rational behavior of economic agents and the bounds of this behavior. In our case, we are interested in the rational behavior of cyberattackers and defenders. Behavioral models typically integrate insights from psychology, neuroscience, and microeconomic theory and, in our case, cybersecurity theory. The study of behavioral economics includes how financial decisions are made.

We have chosen to introduce this topic in a study of the behavioral aspects of cybersecurity because, as we have seen in previous chapters, we can often describe cybersecurity scenarios using an economics model. Among the leading scholars in this new branch of economics are Daniel Kahneman, Amos Tversky, and others. Three Nobel prizes in Economics have been awarded for this field of research in the past decade.

ORIGINS OF BEHAVIORAL ECONOMICS

In the 1960s, cognitive psychology began to shed more light on the brain as an information processing device (in contrast to behaviorist models). Scholars such as Amos Tversky and Daniel Kahneman began to compare their cognitive models of decision-making under risk and uncertainty to economic models of rational behavior.

It is suggested that humans take shortcuts that may lead to suboptimal decision-making.

A seminal work in introducing this field appeared in 1979, when Kahneman and Tversky published *Prospect Theory: An Analysis of Decision under Risk*, which used cognitive psychology to explain various divergences of economic decision-making from neoclassical theory.

UTILITY

Utility is a most interesting word in the English language, with many different connotations. In our case, it will not refer to the backup shortstop on a baseball team, nor the company that provides the distribution of electrical services to a community.

In economics, generally, and behavioral economics, in particular, the concept of utility is used to denote worth or value.

If you have a set of potential economic alternatives and you have a way of ordering these preferences in your favor, a utility function can represent those preferences if it is possible to assign a number to each alternative; then, also, you can say that alternative a is assigned a number greater than alternative b, if and only if you prefer alternative a to alternative b. Consider creating an example that will illustrate the thinking about maximizing utility over the first century-plus of utility theory. Suppose that we value a number of commodities at different levels. For the sake of the example, let us choose pairs of shoes, and let us say that we value them as follows:

Flip-flops	$10
Sandals	$20
Sneakers	$30
Overshoes	$40
Dress shoes	$50

Classical utility theory concludes that if you have a range of choices of different values, and every one of the choices increases in value by the same amount, the set of choices would remain the same. Going to our example in the chart, if a consumer decided that the greatest value to him or her would be the $30 sneakers, and if the shoe store next door to the original one was selling sneakers at the same prices also offering a free $30 gift bag, then the customer would still choose the sneakers among the other choices. In general, we describe the utility function as u( ), i.e. u(Flip-flops) = $10, u(Sandals) = $20.

CHALLENGE TO UTILITY THEORY

Here are two examples developed by Kahneman and Tversky (1979):

We first show that people give greater weight to outcomes that are considered certain relative to outcomes that are merely probable (the certainty effect). Here are two choice problems: The number of respondents who answered each problem is denoted by N, and the percentage of respondents who chose each option is given in brackets.

We will show the paradox demonstrated by these results. We assume that the specific choice chosen more often is indicated by the majority of results for the N subjects. In Table 16.1, 18% of the subjects chose A, and 82% chose B. But for the same subjects, when asked to choose for Table 16.2, 83% of the subjects chose C, and 17% chose D.

Clearly the subjects picked choice B from Table 16.1 by a substantial majority. The same respondents, when given Table 16.2, were even more definitive in choosing choice C over choice D.

TABLE 16.1

Choose between choices A and B Where You Would Receive a Payoff in a Given Amount, Subject to the Corresponding Probability

Choice A		Choice B	
Payoff	Probability	Payoff	Probability
$2500	0.33	$2400	1.00
$2400	0.66		
$0	0.01		

Result of testing the experiment with N = 72 subjects:

18%	72%

TABLE 16.2

Choose Between

Choice C		Choice D	
Payoff	Probability	Payoff	Probability
$2500	0.33	$2400	0.34
$0	0.67	$0	0.66

Result of testing the experiment with N = 72 subjects

83%	17%

This pattern of preferences violates expected utility theory in the manner originally described by Allais.

According to that theory, with $u(0) = 0$, Table 16.1 implies, since choice B is preferred by the subjects over choice A:

$$0.33u(2500) + 0.66u(2400) < u(2400)$$

$$\Rightarrow 0.33u(2500) > 0.34u(2400)$$

while the preference in Table 16.2 implies the reverse inequality:

$$0.33u(2500) < 0.34u(2400)$$

Note that Table 16.2 is obtained from Table 16.1 by eliminating a 0.66 chance of winning $2400 from both prospects under consideration. In other words, the representation of the choices from A and B to C and D is a constant. Classical utility theory would say that since the relationship between the choices has not changed, the preference for the outcome would not change. But this change produces a greater reduction in desirability when it alters the character of the prospect from a sure gain to a probable one than when both the original and the reduced prospects are uncertain. A simpler demonstration of the same phenomenon involving only two-outcome gambles is given in the following. This example is also based on Allais.

Again, according to classic utility theory, with $u(0) = 0$, Table 16.3 implies, since choice B is preferred by the subjects over choice A:

$$0.80u(4000) < u(3000)$$

while the preference in Table 16.4 implies the reverse inequality:

$$0.20u(4000) > 0.25u(3000) \Rightarrow 0.80u(4000) > u(3000)$$

(multiply each side by 4)

In this pair of problems, as well as in all other problem pairs in this section, almost half the respondents (45%) violated expected utility theory by flipping from one side to the other. To show that the modal pattern of preferences in Tables 16.3 and 16.4 is not compatible with the theory, set $u(0) = 0$, and recall that the choice of B implies $u(3000)/u(4000) > 4/5$, whereas the choice of C implies the reverse inequality.

TABLE 16.3
Choose Between

Choice A		Choice B	
Payoff	**Probability**	**Payoff**	**Probability**
$4000	0.80	$3000	1.00
$0	0.20		
Result of testing the experiment with N = 95 subjects			
20%		80%	

TABLE 16.4
Choose Between

Choice A		Choice B	
Payoff	**Probability**	**Payoff**	**Probability**
$4000	0.20	$3000	0.25
$0	0.80	$0	0.75
Result of testing the experiment with N = 95 subjects			
65%		35%	

TABLE 16.5
Choose between choices A and B Where You Would Receive a Nonmonetary Payoff in a Given Amount (Even Though the Subjects Could Perform a Conversion to a Monetary Value), Subject to the Corresponding Probability

Choice A		Choice B	
Payoff	**Probability**	**Payoff**	**Probability**
Three-week tour of England, France, Italy	0.50	One-week tour of England	1.00
No tour	0.50	No tour	0.00
Result of testing the experiment with N = 72 subjects			
22%		78%	

The examples of Tables 16.5 and 16.6 demonstrate that the same effect, the certainty effect, even when the outcomes are nonmonetary.

TABLE 16.6
Choose between choices A and B Where You Would Receive a
Nonmonetary Payoff, Subject to the Corresponding Probability

Choice A		Choice B	
Payoff	**Probability**	**Payoff**	**Probability**
Three-week tour of	0.05	One-week tour	0.10
England, France, Italy		of England	
No tour	0.95	No tour	0.90

Result of testing the experiment with N = 72 subjects
67%   33%

TABLE 16.7
Choose between choices A and B Where You Would Receive a
Payoff in a Given Amount, Subject to the Corresponding
Probability

Choice A		Choice B	
Payoff	**Probability**	**Payoff**	**Probability**
$6000	0.45	$3000	0.90
$0	0.55	$0	0.10
$0	0.01		

Result of testing the experiment with N = 66 subjects
14%   86%

The same effect is also demonstrated when there is a significant
probability of an outcome as opposed to when there is a possibility
of winning, even though extremely remote. Where winning is possible but not probable, most people choose the prospect that offers
the larger gain.

Note that in Table 16.7, the probabilities of winning are substantial (0.90 and 0.45), and most people chose the prospect where
winning is more probable. In Table 16.8, there is a possibility of
winning, although the probabilities of winning are minuscule (0.002
and 0.001) in both prospects. In this situation where winning is possible but not probable, most people chose the prospect that offers the
larger gain.

TABLE 16.8

Choose between choices A and B Where You Would Receive a Payoff in a Given Amount, Subject to the Corresponding Probability

Choice A		Choice B	
Payoff	**Probability**	**Payoff**	**Probability**
$6000	0.001	$3000	0.002
$0	0.999	$0	0.998

Result of testing the experiment with N = 66 subjects

73%	27%

APPLICATION OF THE KAHNEMAN–TVERSKY APPROACH TO CYBERSECURITY

Consider now the development of similar choice problems set in the cybersecurity environment. Suppose we have an attacker who has identified as a target for a ransomware attack a potential victim for which a hacker can assign a monetary value as a ransom.

The attacker can launch a ransomware attack for which he or she could estimate the likelihood of obtaining the ransom. He or she could configure the attack to request a ransom of either $2400 or $2500 and estimates that the lower level of ransom will, on the one hand (choice A), succeed 66% of the time for the $2400 ransom, but only 33% of the time for the larger ransom, and fail completely 1% of the time. Or, the attacker's second choice would be to simply employ the $2400 ransom request, which it is felt would succeed 100% of the time.

Problem 16.1: Choose between choices A and B where you would receive a payoff in a given amount, subject to the corresponding probability.

Choice A		Choice B	
Payoff	**Probability**	**Payoff**	**Probability**
$2500	0.33	$2400	1.00
$2400	0.66		
$0	0.01		

On the other hand, the attacker could make the following choices: Problem 16.2: Choose between:

Choice C		Choice D	
Payoff	**Probability**	**Payoff**	**Probability**
$2500	0.33	$2400	0.34
$0	0.67	$0	0.66

The numbers may seem familiar from the example in Table 16.2, and of course this is not by accident. Although this experiment has not yet been conducted (to our knowledge), if the results would be similar to the Allais–Kahneman–Tversky data, this would be very useful in comparing cybersecurity economic behavior to the more general research on microeconomics in the context of what we are learning in behavioral economics.

NUDGE

Nudging is a subtle (or perhaps "stealth") mechanism to influence behavior as opposed to direct education or legislation. Nudge is a concept in behavioral science, political theory, and economics that proposes positive reinforcement and indirect suggestions as ways to influence the behavior and decision-making of groups or individuals.

- One of the most frequently cited examples of a nudge is the etching of the image of a housefly into the men's room urinals at Amsterdam's Schiphol Airport, which is intended to "improve the aim." Men now had something to aim for—even subconsciously—and spillage is reduced by 80%.
- Nudge theory urged healthier food be placed at sight level in order to increase the likelihood that a person will opt for that choice instead of less healthy option.
- In some schools, the cafeteria lines are carefully laid out to display healthier foods to the students. In an experiment to determine its effect, it was shown that students in the healthy lines make better food choices, with sales of healthy food increasing by 18%.

Both Prime Minister David Cameron and President Barack Obama sought to employ nudge theory to advance domestic policy goals during their terms.

AN APPLICATION OF NUDGE THEORY TO CYBERSECURITY

Suppose there are ten files, each containing some information of value, perhaps, for example, the location of a bank account. All of these files are encrypted on the host's system. A cyberattacker is able to penetrate the system and determines that he or she has enough time before being detected to obtain the information from at least one of the ten encrypted files. For 9 of these 10 values, the file sizes are between 1 and 5 kB.

The tenth file, as established by the defender, is information of an economic value of only, say, $5.00. But on the other hand, the file is padded before printing so that the file size becomes 25 kB.

When the attacker finds the 10 encrypted files, he or she may be nudged toward the 10th file, of 25 kB, because the clear perception of the file size being greater than all the others would suggest the potential of greater value in successfully attacking that file. Of course, being nudged toward the 25 kB file will cause the attacker to have to spend considerably more effort to decrypt the file, thus increasing exposure to detection and capture.

PROBLEMS

1. Find your own sample of students to conduct the four Kahneman–Tversky choice problems. Compare your results to the published data.
2. Should one expect the aforementioned results to change depending on the sector of the economy utilized, for example, the cyberenvironment, for one case?
3. Find a number of examples of the application of nudge theory.

REFERENCES

Kahneman, D. & Tversky, A. 1979. Prospect theory: An analysis of decision under risk. *Econometrica. The Econometric Society*, 47(2), 263–291. doi: 10.2307/1914185. JSTOR 1914185.

Wikipedia. 2018. Behavioral Economics. https://en.wikipedia.org/wiki/Behavioral_economics.

17

FAKE NEWS

Clearly, to date, the most publicized set of examples of "fake news" deal with the multiple events that have occurred over a number of years; and, most recently, the events leading to and since the United States 2016 presidential election.

There should be an increasing focus on strategies to identify the intrusions that we might label fake news and the development of techniques to identify and thus defeat such practices. Corporations such as Facebook and Yahoo have been investing in research to address these problems.

However, because there is no clear set of methodologies that can be employed to eliminate this problem, we will try in this section to identify some partial approaches to determining the validity of news and to suggest certain measures in order to identify fake news.

A FAKE NEWS HISTORY

During the period of slavery in the United States, its supporters developed many fake news stories about African-Americans, purported slave rebellions, or stories of African-Americans spontaneously turning white, which brought fear to many whites (Theobald, 2006).

At the end of the nineteenth century, led by Joseph Pulitzer and other publishers and usually referred to as "yellow journalism," writers pushed stories that led the United States into the Spanish-American War when the USS Maine exploded in the Havana harbor (Soll, 2016).

In 1938, the radio drama program "The Mercury Theater on the Air," directed by Orson Welles, aired an episode called "War of the Worlds," simulating news reports of an invasion of aliens in New Jersey. Before the program had begun, listeners were informed that this was just a dramatization. However, most listeners missed that part and therefore believed that the invasion was real. One concrete result was that the attack, supposedly in Grover Mill, New Jersey, resulted in residents attacking a water tower because the broadcaster identified it as alien (Chilton, 2016).

FAKE NEWS RESURGENCE, ACCELERATION, AND ELECTIONS

Over the past decade, the use of fake news has been applied in a number of areas: for financial gain, for political purposes, for amusement, and for many other reasons.

There has now been a great deal written about the use of such fake news in order to influence not only the 2016 US presidential election but also elections in numerous other countries throughout the world. One especially egregious example was the result of the news story usually called the "Pizzagate" conspiracy theory, which accused a certain pizzeria in Washington, DC, as hosting a pedophile ring run by the Democratic Party. In December 2016, an armed North Carolina man, Edgar Welch, traveled to Washington, DC, and opened fire at the Comet Ping Pong pizzeria identified in the Pizzagate fake news. Welch pleaded guilty to charges and was sentenced to 4 years in prison (Kang, 2016).

WHAT IS FAKE NEWS?

There is nothing new about fake news. Most of us first encounter this phenomenon the first time we discover a lie. However, it can fairly be argued that the telling of a lie presented as part of the technology may seem so impressive as to convince us of the truthfulness of the message because of the elaborate wrapping.

It is really only in the Internet age that the toolset has become readily available to create a very (seemingly) realistic message that will fool many readers.

The very rapid expansion in the past few years of fake news sites can be attributed to many factors, for example, the use of composition software to produce websites with very sophisticated appearances provides the opportunity to create such sites.

The motivation for creating such sites is undoubtedly a reason for the rapid expansion. There is a financial incentive for many of these sites, either because their creators are being paid to produce them or because the site itself may provide a mechanism for readers to buy—whether a legitimate product or a scam.

And many sites may have as a greater purpose the advocacy of a political point of view, which may put forward completely false information.

SATIRE OR FAKE NEWS?

A difficult challenge for the consumers of various information is that there may be substantial similarity between sites that might be considered fake news and those that are meant to be satirical. Certainly, satire is an important form of storytelling and criticism.

Indeed, it is well known that many nursery rhymes that we know from childhood are actually satire disguised as children's tales to avoid retribution since the satirical meaning may actually constitute criticism of a powerful monarch. Consider (Fallon, 2014):

LITTLE JACK HORNER

Little Jack Horner
Sat in the corner,
Eating a Christmas pie;
He put in his thumb,
And pulled out a plum,
And said, "What a good boy am I!"

One interpretation has "Little Jack" standing in for Thomas Horner, a steward who was deputized to deliver a large pie to Henry VIII concealing deeds to a number of manors, a bribe from a Catholic abbot to save his monastery from the king's anti-Catholic crusade. It is believed by many that Horner reached into the pie and helped himself to a deed.

YANKEE DOODLE

Yankee Doodle went to town
Riding on a pony
Stuck a feather in his cap
And called it macaroni!

Yankee Doodle is a silly figure in this classic ditty, which dates back to the Revolutionary era. At the time, macaroni was the favored food of London dandies, and the word had come to refer to the height of fashion. British soldiers, who originally sang the verse, were insulting American colonists by implying they were such hicks they thought putting feathers in their hats made them as stylish as London socialites.

There is a major effort now from many quarters in trying to identify techniques to be able to classify websites for social media in terms of their "fakeness." The major companies hosting social media, for example, Facebook and Yahoo, have initiatives to identify and disqualify fake news accounts.

DISTINGUISHING SATIRE FROM FAKE NEWS

In the next section, we will identify a number of tests that can be applied to assist in determining the status of a questionable website or Facebook message we might encounter. There has been an explosion in the number of fake news sites, but we will analyze a few examples to try to determine their validity. What follows are set of examples of either "fake news" or some we might designate "not fake news."

DailyBuzzLive.com

DailyBuzzLive is an online magazine that specializes in sensational articles. The flavor can be ascertained from some of the sections of the zine indicated by the menu selections: "Controversial," "Viral Videos," "Weird," "Criminal," "Bad Breeds."

One immediate clue as to its validity is that there is no obvious way of determining the publisher. The topics vary widely, but a few sample headlines include the following:

"USDA Allows US to be Overrun With Contaminated Chicken from China"
"Human Meat Found In McDonald's Meat Factory"
"People Call for Father Christmas to be Renamed 'Person Christmas'"

With respect to the last article regarding Christmas, there is no author indicated, nor any date on the article. Furthermore, the headline, beginning with "People Call for ..." never refers to anyone actually making that "call."

ABCnews.com.co

This website ABCnews.com.co is no longer in existence. If you enter that URL, you will find the statement that is common to nonexistent sites that begin with "Related Links." This one is easy to detect: the URL, although appearing to be the website for ABC News (ABCnews.com), actually ends with the Internet country code ".co" for Colombia.

TheOnion.com

On first glance, one might consider *The Onion* fake news. Although not doing so on the masthead, *The Onion* identifies itself as satire, and it follows a lengthy tradition in this genre, with a political impact. *The Onion* is unlike the *Daily Buzz* example above, which does not so identify itself. Several of the Onion articles, clearly satirical, are as follows:

"Nation Not Sure How Many Ex-Trump Staffers It Can Safely Absorb"
"New Ted Cruz Attack Ad Declares Beto O'Rourke Too Good for Texas"
"Elizabeth Warren Disappointed After DNA Test Shows Zero Trace of Presidential Material"

The Onion on its website indicates satirically the supposed history of its publication:

> *The Onion* is the world's leading news publication, offering highly acclaimed, universally revered coverage of breaking national, international, and local news events. Rising from its humble beginnings as a print newspaper in 1756, *The Onion* now enjoys a daily readership of 4.3 trillion and has grown into the single most powerful and influential organization in human history.

Infowars.com

This site is closely affiliated with Alex Jones, who has long been identified as a conspiracy theorist, ranging from such conspiracies as the "birther conspiracy" about former President Obama to the argument that school shootings such as Sandy Hook and Lakeland were faked, to the "Pizzagate" story cited earlier in this chapter.

However, beyond that connection, the other giveaway is the headline photograph of what is entitled "Invasion Begins! Migrant Caravan Arrives At US/Mexico Border" showing immigrants scaling a wall, presumably to enter the United States (https://www.infowars.com/invasion-begins-migrant-caravan-arrives-at-us-mexico-border/). When one checks other media sources, at the time this article was published (November 15, 2018), many reports from other sources indicated that the "caravan" of refugees (so designated by Donald Trump during the 2018 election campaign) was hundreds of miles from the US border.

New Yorker

Andy Borowitz has written a series of satirical articles for several years in the New Yorker (https://www.newyorker.com/contributors/andy-borowitz). He makes it very clear in the headline over his articles that they are "Satire from the Borowitz Report"; in addition, also above his articles is a line indicating they are "Not the News." Borowitz created the Borowitz Report in 2001

Empirenews.net

In one alarming story, the person indicated in the headline, Jerry Richards, is alleged to have murdered over 700 people in Naples, Florida. In this case, since allegedly the murderer has been arrested, there should be court records as well as other news articles about this event. There are not.

At least, in an "About" section, *Empire News* indicates that "it is intended for entertainment purposes only."

Beforeitsnews.com

The November 12, 2018, article "Operation Torch California" (https://beforeitsnews.com/v3/terrorism/2018/2461448.html) regarding the devastating forest fire both in Northern and Southern California raised a number of questions. The first warning sign occurred with the quote in the opening: "*Operation Torch California* is a very real ongoing black operation being conducted by the U.S. Intelligence Community ... first and foremost a highly sophisticated psyop." The quote was attributed to an unnamed "Intelligence Analyst & Former

U.S. Military Officer." Next, there are references to the acronym DEWs, which is never defined.

From that point on, the article bounces from one wild statement to another, linking this story with Hurricane Michael in Florida, to ISIS and Al-Qaeda, to aluminum oxide from coal fly ash. Among other wild statements, it is alleged that "To name the most devastating fire in California history *Camp Fire* represents the profound cynicism associated with this well-planned pyro-psyop. How easy it is to now blame that geoengineered wildfire on a simple 'camp fire'."

Centers for Disease Control

https://www.cdc.gov/phpr/zombie/index.htm

This is perhaps the most perplexing of all the examples, since it appears on the website of an agency of the US government, the Centers for Disease Control (CDC) based in Atlanta (Figure 17.1).

The CDC has as its charge the battle against infectious diseases, and it usually is at the forefront when there are outbreaks such as the Zika virus, Ebola virus, or the coronavirus. However, to advise people about how to deal with a "Zombie Apocalypse" seems to be unusual, to say the least. Will readers actually believe that it is necessary to prepare for such an event? Or will everyone realize that

Figure 17.1   CDC "Zombie Apocalypse."

this is merely satire intended to heighten concern generally about how infectious disease can spread? Unfortunately, the CDC, for whatever reason, does not choose to identify the site as satire.

ASSESSING FAKE (OR NOT-FAKE) NEWS

It may be instructive to see what can be learned from human intervention. We will look at a number of these sites to see what "tells" or techniques we can use to identify them as fake.

Fake News Detecting Technique	Explanation
Author bibliography	See what you can find out about any author indicated. Does that person exist? If so, what are his or her credentials or bibliography?
Authorless	Be suspicious if an article appears and no author is credited.
Comment section	If the website has a comment section, see if you can determine the credibility and nature of the comments.
Emotional reaction	How do you feel about the story? Or perhaps, "does it pass the smell test?" Your reaction might depend upon the content.
Expert testimony	If any experts are quoted, search to see if they really exist; or, if they do, what are their qualifications.
Fact checkers	There are a number of fact-checking organizations that can be consulted to see if the news item is legitimate. A list of these fact checkers will follow.
Grammar	Look for spelling and punctuation errors. It may help to copy the text into Word and run the spellchecker there.
Included Ads	Examine the nature of the ads that might be featured on the suspected site. If the ad indicates you can purchase something online, it may be a scam.
News outlet	If a news outlet is indicated, and you have not heard of it, search online for more information.
Other articles	See if there are other articles on the same topic. If you cannot find any, the chances are the story is fake.

(Continued)

Fake News Detecting Technique	Explanation
Other sources	Do a search to see if the story at hand is also covered by other media. If it does not appear in the same time frame in a reputable medium, it may very well be bogus
Publisher "About Us"	In the "About Us" section of the website, see what you can determine about the organization sponsoring the site.
Purpose of the story	Try to determine the purpose of the story. Is it possible that it is to satisfy an agenda of the publisher, for political or financial reasons?
Quotes	If the quote is given, search for the source of the quote. See also if you can determine if the person being quoted actually exists or has actual credentials.
Reverse image search	If you right-click on an image, you will find an option to search for the image. Then you should be able to see other websites that have used it and if they are relevant.
Seek other experts	If someone in an article is cited as an expert, see how that person is considered by other experts in the same field.
Source check	See if the publisher meets academic citation standards.
Timeliness	Can you verify if the article is recent or perhaps is a copy of something written years before?
URL	When you access the site, examine the URL carefully. On occasion, fake sites have acquired URLs in unlikely countries.
Visual assessment	Just consider the overall appearance of the site. Once again, you may be able to apply a "smell test."

When seeking fact-checking organizations, you may try these:

FactCheck.org (http://www.factcheck.org)
Politifact (http://www.politifact.org)
The International Fact-Checking Network (https://www.poynter.org/channels/fact-checking)
Snopes.com (http://snopes.com)

PROBLEMS

1. Find sources (necessarily an octogenarian+) with a personal recollection of the "War of the Worlds" radio broadcast in 1938. Summarize their recollections.
2. Find any recent (post-2016) reference to "Pizzagate."
3. We have provided eight examples of fake news (or not-fake-news) above. Submit each to the battery of 20 tests indicated in "Fake News Detecting Techniques." What metric would you use to make a final determination of fake news versus not-fake-news? For example, you might say it is fake news if it meets >n of the 20 tests.

REFERENCES

Fallon, C. 2014. "The Shocking, Twisted Stories behind Your Favorite Nursery Rhymes," *Huffington Post*, November 20. https://www.huffingtonpost.com/2014/11/20/nursery-rhymes-real-stories.

Kang, C. 2016. "Fake News Onslaught Targets Pizzeria as Nest of Child-Trafficking," *The New York Times*, November 21. https://www.nytimes.com/2016/11/21/technology/fact-check-this-pizzeria-is-not-a-child-trafficking-site.html.

Soll, J. 2016. "The Long and Brutal History of Fake News," *POLITICO Magazine*, December 18. http://www.politico.com/magazine/story/2016/12/fake-news-history-long-violent-214535.

Theobald, M. M. 2006. Slave conspiracies in Colonial Virginia. *Colonial Williamsburg Journal*. http://www.history.org/foundation/journal/winter05-06/conspiracy.cfm. http://www.history.org.

18

EXERCISES: HACK LABS

The purpose for this lab is that it is not necessary, but there could be a supportive physical computer lab to carry out these projects. They can also be done on a student's own computing equipment and do not have to be done within a fixed lab period.

When these have been offered by the authors, they have usually allowed the students a week to carry out the research and submit the results.

HACK LAB 1: SOCIAL ENGINEERING: FIND COOKIE'S PASSWORD

The first of these labs deals with an approach used by many hackers that is sometimes called social engineering or dumpster diving. Most naïve or even not-so-naïve computer users do not realize how an attacker may try to gather sufficient information about a target in order to be able to make an educated guess about the target's password.

It should also be remembered by those who are responsible for maintaining security on a multiuser system, such as a university's primary network, that an attacker does not need to capture thousands of passwords in such a system: Only one will probably suffice to meet the hacker's need.

This lab will ask students to try to determine Cookie Lavagetto's password. It is given that Cookie, as a somewhat careless user, has a password that is constructed from two separate pieces of information about Cookie and his family. In the data sheet that follows, note that there are 21 distinct data points regarding Cookie and his family: family names, addresses, children, their ages and birthdays, their pets, and pet names. This is information that most dumpster divers should be able to accumulate.

Incidentally, the inspiration for this example is from the best film ever about computer hacking, the 1983 thriller called *War Games* where the protagonist is a teenager who is trying to find a password to break into a gaming computer. According to the plot, it is actually a defense department computer that could start World War III invented by a computer scientist named Falken. How the teenager discovers Falken's secret password is not only a fascinating part of the film but also a marvelous lesson in the techniques of password finding or dumpster diving.

With 21 data points and a rule established that a password can be no more than a combination of two of the given data points, there are $(21 \times 20)/2 = 210$ possibilities. You might limit the lab to trying, say, 30 combinations.

Cookie's Dataset: Instructions to Student
  Cookie Lavagetto
  Spouse: Cookette Lavagetto
  Residence: 200 Ebbetts Field Road, Brooklyn, NY 10093
  Children: Oreo Lavagetto, age 14; Gingersnap Lavagetto, age 12; Tollhouse Lavagetto, age 7
  Pets: Spot, Woofie, George
  Birthdays: Cookie 1/23, Cookette 8/17, Oreo 3/9, Gingersnap 11/28, Tollhouse 4/3

Cookie's email: The instructor creates this.
  How to play: Log in to the given account.
  Try to guess a password. Suggestion: Before entering electronically, construct and write down 10 guesses. Then see if you have guessed the correct password.
  The correct password is some combination of the data above. It will be no more than 14 characters.
  If you are successful in guessing the password, you have hacked Cookie's account. Submit as the assignment (by email) the successful password you have entered. *Like any good hacker, keep your successful password to yourself.*
  The Hack Lab will run for 1 week or until you make 30 attempts. I will change the password three times during the week, so you may have a total of four correct solutions. You will get full marks if you get any two of the four.

HACK LAB 2: ASSIGNED PASSWORDS IN THE CLEAR

This Hack Lab deals with ways to discourage online companies from requiring a user to create a password to avail himself or herself of the company services. Most of us who are frequent users of online services find that they may have to manage many such passwords. However, a number of companies, presumably with little concern for security, have the habit of requiring the user to obtain a password by signing up and then having it emailed to the user's account and displaying the host-created password "in the clear."

Instructions:

a. Find a (or many) website(s) for which you must create a password, and then the password is sent to you in email *in the clear*!

b. No responses categorized in plaintextoffenders.com will be accepted. Deadline: 1 week after lab announcement, but this site may give you some good ideas.

HACK LAB 3: SWEENEY PRIVACY STUDY

The purpose of this hack is to confirm the research of LaTanya Sweeney at Harvard (Sweeney, 2000). Her research showed definitively that 87% of the US population can be uniquely identified by three commonly available data points: {birthdate of the individual including year, gender, and US zip code (5-digit version) of residence}.

A test bed of 20 data sets is provided as samples for carrying out this lab. The objective for students is to try to identify the individuals for whom, in most cases, sufficient information is provided to determine an individual in the United States who fits the given data regarding birth date and zip code of residence.

When this experiment was conducted by our students at Howard University, on separate occasions, usually about half the class would identify 10 individuals given their personal data within 24 hours.

The only additional advice given to students was that the subjects to find had some measure of celebrity.

Dr. LaTanya Sweeney, now professor of government and technology at Harvard, published in her PhD thesis in computer science at MIT that it was possible to identify uniquely over 85% of the US population with only three pieces of easily available information: the gender, the exact birth date of the individual, and the postal zip code of their address. With this information, and with general access to health records without names being protected or encrypted in a health database, a health researcher could construct a legitimate query that would result in only one record being returned, and thus by independently discovering the information described earlier (exact birth date and zip code), discover, without hacking or illegal access, the health records of an individual. This demonstrates a problem in maintaining secure databases.

In fact, Dr. Sweeney was able to discover that the then governor of Massachusetts had not disclosed health information that could have affected his performance in public office as governor.

In this assignment, I have provided the exact birth dates and zip codes of ten real people, using publicly available information and constructed (with *fake* diseases) the results of a query or queries into a health database that returned the following (names being obscured or suppressed):

Name	Birth Month	Birth Day	Birth Year	Zip Code	Disease
	November	14	1954	94306	Kidney disease
	March	1	1994	90068	Pneumonia
	June	8	1925	04046	Arthritis
	February	20	1963	85258	HIV
	June	1	1937	38921	Migraine headaches
	August	30	1930	68132	Ruptured spleen
	July	31	1958	75220	Alzheimer's disease
	August	11	1965	91344	Gout
	March	25	1942	48302	Chronic bronchitis
	October	28	1955	98112	Macular degeneration

Your assignment is to use this information to discover the individual in each case. Like any good hacker, you will not discuss your work with anyone else—after all, the person you might discuss this with might turn you into authorities. For each name you discover, email that person's name and the disease to me by email. You get 1 point (out of 10) for each correct solution.

HACK LAB 4: PASSWORD METERS

This Hack Lab is meant to have students judge the consistency and effectiveness of a number of password meters that can be found on the Internet.

Password Meter Designation	Name
A	https://passwordmeter.com
B	https://lastpass.com/howsecure.php
C	https://my1login.com/resources/password-strength-test/
D	https://thycotic.com/resources/password-strength-checker/
E	https://howsecureismypassword.net/

The following is one sample result using sample passwords generated by the authors.

The formula that can give you a single datum comparing all solutions would be as follows:

Suppose the indices we use are the indices in the Test Password array. For the aforementioned example, take the square of the differences of all elements in each row, and call the total DISTANCE. The columns are called A, B, C, D, E and the rows 1, 2, 3, 4, 5, 6:

$$\text{DISTANCE}i = (Bi - Ai)^2 + (Ci - Ai)^2 + (Di - Ai)^2$$

$$+ (Ei - Ai)^2 + (Ci - Bi)^2 + (Di - Bi)^2$$

$$+ (Ei - Bi)^2 + (Di - Ci)^2 + (Ei - Ci)^2$$

$$+ (Ei - Di)^2$$

Repeat this computation for i = 1, 2, 3, 4, 5, 6.

Showing only the calculation for DISTANCE1:

$$(2-6)^2 + (6-6)^2 + (3-6)^2 + (1-6)^2 + (6-2)^2 + (3-2)^2$$

$$+ (1-2)^2 + (3-6)^2 + (1-6)^2 + (1-3)^2$$

$$= 16 + 0 + 9 + 25 + 16 + 1 + 1 + 9 + 25 + 4 = 106.$$

PROBLEMS

(Hack Lab 1)

1. Cookie has 21 data points. If a password is constructed from a combination of two data points, how many tries will it take to guess the correct password?

Test Password	A	B	C	D	E
11111111111111111111	6 Very weak 0%	2 Weak	6 Very weak 0.04 seconds	3 2 weeks	1 79 years
Penob7scot	2 Strong 63%	2 Moderately strong	2 Strong 7 months	1 3 years	2 8 months
x3p9q!m	3 Good 54%	4 Weak	1 Strong 8 months	6 1 minute	6 22 seconds
brittttany	5 Very weak 8%	2 Moderately strong	5 Weak 16.05 minutes	5 4 hours	5 59 minutes
Onomatopoeia	4 Very weak 13%	1 Very strong	3 Medium 13 hours	2 4 months	3 4 weeks
aBc123xYz	1 Strong 76%	4 Weak	4 Weak 40.01 minutes	3 2 weeks	4 4 days

2. Construct a separate example. Allow a classmate five guesses. How many of your classmates will find the password?
3. Can you find a real-world example of dumpster diving?
4. How did the high school student in the movie *War Games* discover Falken's password? What was this password and why would Falken have chosen it?

(Hack Lab 2)

1. How would you prevent exposure of your password if it is delivered from a website in the clear?

(Hack Lab 3)

1. Carry out this Hack Lab, first for the first table above, then for the second table. You will be above the 50th percentile of our students if you complete either table within 1 day.
2. Create your own set of 10 dates and zip codes, and exchange with a classmate.
3. Give an estimate of Sweeney's figure of 87% individual identification for:
   a. Canada
   b. United Kingdom
   c. Brazil

(Hack Lab 4)

1. Use your own password choices to test the DISTANCE algorithm. See if you can find password choices that maximize the distance function described earlier.

REFERENCES

Plaintext Offenders. 2018. http://plaintextoffenders.com/.
Sweeney, L. 2000. *Simple Demographics Often Identify People Uniquely. Data Privacy Working Paper 3.* Carnegie Mellon University, Pittsburgh, PA.

19

CONCLUSIONS

You have now arrived at the end of our story—for now. We hope you will find it an interesting challenge to carry on your learning about these increasingly important issues arising from cybersecurity and the behavioral sciences. We believe this compound subject involving both the knowledge of computing disciplines and behavioral science disciplines is still in its infancy. In particular, there are paths that can lead to important future research understanding based on the following.

PROFILING

We have provided a few examples of profiling techniques, but the reader can find many other examples of accounts of other attacks that have taken place in recent years. The Sony Pictures example we have used is interesting because it touches on a number of varieties of subjects and motivations. But the reader might find it equally of interest to analyze the cases of WannaCry, Petya, fake news incidents, and numerous others.

SOCIAL ENGINEERING

Social engineering provides a continuing and important avenue for the development of hacking approaches to unlawful entry. A student studying the success of social engineering methods can approach it from two perspectives: first, understanding the exposure that a careless user provides to a hacker or "dumpster diver" in trying to guess a certain password. On the other hand, one can also be cognizant of the security provided when password methods are developed with great care. This book has also provided an introduction to the utility or futility of what we now call password meters, and the field is open for persons who will potentially design criteria for such meters in a more rigorous fashion.

Persons interested in this area of study should also take into consideration the challenges of trying to move from what we have referred to as one-factor authentication, such as passwords or biometrics, to multifactor authentication, both from the perspective of the security provided but also the difficulty in migrating from the world's supply of password system users to a multifactor environment.

SWEENEY PRIVACY

The results of Sweeney's research (Chapter 7) showing that almost 90% of users in the United States can be determined uniquely by three usually widely available pieces of information are shocking to many people. Further research in this area might consider the level of security provided by other forms of identification. Future research by the authors of this book will illustrate that the data described in Sweeney's research make the issue of privacy even less secure in a number of countries, while similar data do not pose as much of a threat in certain other countries.

UNDERSTANDING HACKERS

We have tried to provide a number of examples of the psychology of certain hackers that have been willing to describe themselves in the literature. But this is a field that is wide open to further study. There has not been a great deal of literature on the analysis of motivations of a greater number of identified hackers and even less on the understanding of behavioral tendencies of hacker organizations.

GAME THEORY APPLICATION TO PROFILING

Game theory has a considerable body of scholarship in general but has had very little application to the field of cybersecurity. We believe that in all probability this is because researchers in cybersecurity problems have not tended to take a quantitative approach to the research. This is an approach we have tried to emphasize throughout this book, and although we realize that in a number of cases attempting to quantify issues related to behavior can be challenging, we feel that this is not the reason to avoid a quantitative approach altogether, such as we might use in game theory techniques.

TURING TESTS

The Turing test remains a fascinating example, bequeathed to us by Alan Turing, of a methodology for distinguishing not only between human and computer but also, as we have projected, between different groups of humans (or machines) in our environment. The specific example we have pursued is related to the determination of gender, as we have called it, the "gender Turing test." But readers may also be interested in developing an "age Turing test," a "professional training Turing test," or conceivably others.

CRYPTO AND STEGO

An area that could bear very significant research is the concept of developing security measures that use a hybrid approach involving both cryptography and steganography. It would seem that the importance of developing such an approach would be to use the aspect of steganography that tests human behavior as a tool supplemental to that provided by cryptography.

BEHAVIORAL ECONOMICS

This field has become a major area of research in economics, although many would also argue that it is as significant a subdiscipline in psychology as it is in economics. In our case, there has been very little study about how this may be applied to issues of cybersecurity, so this field is in our minds wide open.

FAKE NEWS

This is in many ways a new area on the scene. Although, as we described in Chapter 17, we can track "fake news" virtually to the dawn of human history, the distribution of such news has of course been empowered by our current technological area, in particular with the use of the Internet. The question that we have suggested for further research is the development of human-developed or artificial intelligence techniques for the detection of what we call fake news, and this is not only a wide-open area but also of substantial interest to major organizations in the computing industry.

PASSWORD METERS

We have mentioned the, generally speaking, lack of success in developing meters to measure password strength. There is a question as to whether this field can be improved or whether there are standard measures that could apply to all situations to determine password strength or the lack of it.

NEXT STEPS

The authors are very interested in encouraging the development of coursework and research in this emerging field of behavioral cybersecurity. As a consequence, we will be very pleased to hear from many students and professors involved with either coursework or research in this new and promising area.

We encourage contact of any form and suggest that the best means of informing us of your progress is the email address:

waynep97@gmail.com
Wayne Patterson
Cynthia Winston-Proctor
Washington, DC
August 2020

INDEX

ABCD approach 40, 42–44, 137
access control 45
Access 1, 4, 8, 12, 30–31, 45, 47,
     52–57, 156, 160–161
Adelman, Len 121
AES (Advanced Encryption
     Standard) 102, 108–112,
     119, 122–123
algorithm 49, 59, 64, 108–109,
     110, 120, 121, 127, 129,
     134, 136
Allais, Maurice 142, 146
Anonymous 2–3, 38–39, 56
arithmetic modulo n 60, 98, 101
Art of Intrusion 28
asymmetric encryption 119
attacker/defender scenario 133
authentication 45–46, 49, 50, 52,
     120, 166
authorization 8, 45, 49, 52–53

Beforeitsnews.com 153
behavioral economics 13, 14,
     138–139, 146
BLP (Bell-LaPadula model) 52, 53,
     57–58
biometrics 46, 49
Bitcoin 5
Bletchley Park 66, 88
Borowitz, Andy 153
Bot 133
Buss, David 80

Caesar, Juliusi 59–60, 62–63, 66
Centers for Disease Control 154
ciphertext 59–60, 66, 109, 120,
     124, 127, 134
COVID–19 (Coronavirus) 10
Clinton, Hillary 6–7, 126
covert channel 54–55, 57, 124
cyber crime 9, 15
cyberattack 8, 14–15, 35, 41, 86,
     133, 139, 147
cybersecurity
     access control 54, 57
     authorization 8, 45, 49, 52–53
     behavioral economics 13–14,
          139–140, 146
     behavioral science 13, 32,
          123, 164
     cryptography (historic) 59
     cryptography (modern) 100,
          107, 124, 129, 131
     ethical hacking 28, 31, 32
     fake news 10, 148–149,
          151–152, 155, 167
     game theory 40, 69–70,
          79, 137
     hack labs 153, 157, 158
     hackers 28–33
     profiling 35, 37, 40–41
     steganography 124–125, 127,
          129, 131, 167
     turing test 86–87, 90–91,
          95–96, 167
cyclic alphabet 60

Daemen, Joan 108, 112, 117, 123
DailyBuzzLive 151
DDoS (Distributed Denial of
      Service) 2–4, 11,
      42–43
decryption 65, 97, 112, 120, 134
Department of Justice 8, 9
DES (Data Encryption Standard)
      108

encryption 61–65, 88, 95, 97,
      103, 107–108, 110, 112,
      119, 121–122, 130, 134,
      136–137
Enigma machine 66–67, 88, 90
equalizing strategy 71, 74, 79
ethical hacking 28, 31, 32, 34
exhaustive search 1, 61

Facebook 7, 10, 148, 151
FactCheck.org 156
factoring 101, 134–137
fake news 10, 148–152, 155, 165
Fermat, Pierre de 101, 111
fingerprint 2, 46, 49
finite field 72, 97–98, 101–102, 109
firewall 30, 46, 56

Galois fields 102, 104, 109,
      111, 116
Galois, Evariste 102, 106,
      109, 116
GCD (greatest common divisor)
      101, 104
gender Turing test 86, 96
GOA (Government Accountability
      Office) 1
God's Apostles 36
GoP (Guardians of Peace) 36, 39, 78
group 3, 5, 21, 39, 53, 146, 167

hackers 28–32
Hilton, Peter J. 88, 96
Hoover, J.Edgar 126, 132
Howard University 4

identity role or social identity
      and social structural
      approach 13, 16–17,
      22–27, 81–85, 95, 98, 117
Image File Formats 46, 125,
      127–131, 146, 156
The Imitation Game 88, 90–91
inference 55, 57
Infowars 152
integers 97–98, 101–102
The Interview 36–38
Internet 1, 3–4, 11, 28, 43, 47, 54,
      88, 94–95, 129, 149, 152,
      161, 167
Internet Worm 1
intrusion detection 28, 33–34
inverse to the key 60–61, 99–100,
      103, 111, 113, 122

Jeopardy! 90, 95–96

Kahneman, Daniel 139–141,
      146–147
Kaspersky Labs 6
Kerckhoff's principle 107
keyword mixed alphabet 61
Kittab, Waddah 129, 132
Krawetz, Neil 94, 96

linear algebra 72, 75–77, 103, 135
Little Fermat Theorem 150
lossless compression 131
lossy compression 127–128, 131
Low Orbit Ion Cannon 4, 42
Lu, Donna 32, 34

malware 42–43
Manning, Bradley/Chelsea 2–3
matrix algebra 37, 40–41, 50,
      69–70, 72–76, 78, 93,
      102–103, 111, 116–118
matrix game 73, 76
Matryoshka steganography 129, 132
Mercury Theater 148

methodologies 45, 148
methods 8, 12, 16, 59–60, 75, 88,
          97, 124, 126, 129, 131, 165
metric 133, 135–136
Michael Jackson phenomenon 3–4,
          133, 154
Miller, George 62
minimax strategy 72
Mitnick, Kevin 28, 34
Möbius 100, 104
modular arithmetic 62, 97, 100
Morris, Robert 1, 10, 44, 69, 79
Mossad 6
motivation 17, 20–22, 25–27,
          38–40, 77, 82, 149
multilevel security models 52, 54
Munson, Lee 30, 33–34
Myers-Briggs types 19, 30, 33
narrative design 13, 16–17, 22–24,
          25–27, 33
narrative identity 13, 16–17,
          22–24, 33

NIST (National Institute
          of Standards and
          Technology) 108
NSA (National Security Agency) 88
Nauert, Rick 31, 34
neuroscience 139
Nobel Memorial Prize in
          Economic Sciences 139
nudge 146–147

Obama, Barack 6, 35, 126, 147, 152
objects 52–53, 57
Office of Personnel Management
          1–2, 11
The Onion 152
optimal strategy 71, 73, 78

partial and total order 129, 148
password meters 47, 161, 165
password 5, 30, 45–50, 158–162,
          164–166, 168
payoff 141, 143–146

PayPal 2–3, 5–6, 10
Pershing 126
personality traits 13–14, 16–18, 23,
          30, 81–82
personality 13–14, 16–18, 20–23,
          25–28, 69, 79, 85,
          97, 107
Petya 7, 165
phishing 5–7, 10–11
Pingan, Yu 8
Pizzagate 149, 152, 157
plaintext 164
plaintext offenders 160, 164
Playfair Square 64, 67–68
Podesta, John 7, 10
Politifact 155
presidential election 6, 148
privacy 15, 26, 58, 164
profiling matrix 40–41
profiling 35, 37, 40, 44, 165
psychology 6–7, 13–14, 17, 20,
          25–27, 32, 55, 80–85,
          139–140, 166–167
public key cryptosystem 1–3, 10,
          14, 33, 37, 107–108,
          119–121, 123, 127, 133,
          135–136, 138, 161
pure strategy 71–73, 78–79

QuickStego 128, 131–132

random strings 5, 48, 56, 70–71
randomization 63
ransomware 5, 7, 9–11, 43, 145
rational numbers 97, 102, 136, 139
Raymond, Eric Stephen 30, 34
Rijmen, Vincent 108, 112, 117, 123
Rijndael 102, 108–110, 112,
          122–123
Rivest, Ron 120, 121, 123, 138
Roose, Kevin 32, 34
rotor machines 66
RSA Public Key Cryptosystem 51,
          121, 123, 129, 133–134,
          136–137

saddle point 73–74
Scherbius, Arthur 66, 68
secrecy 68
secret key cryptography 59–61,
        120–121, 125, 127,
        129, 159
security levels 5, 9, 15, 29–31,
        33–34, 49–50, 52–55, 78,
        88, 131, 134–135, 158,
        160, 165–167
self-defining memories 18–19,
        21–23, 26, 85
sex differences approach 80–82,
        84–85, 91
Shamir, Adi 121, 123, 134, 138
Shannon, Claude 63, 68, 107
Simon, Herbert A. 28, 34
smart card 46, 49
Smith, Adam 21, 26–27, 85
Snopes.com 156
social engineering 8, 16–17, 19–21,
        36, 81–85, 139, 151,
        158, 165
social motivation 8, 16–17, 19–21,
        36, 81–85, 139, 151,
        158, 165
Sone, Michael Ekonde 13, 15,
        133, 138
Sony Pictures hack 35–39, 66, 68,
        77–79, 88, 96, 165
steganography 124–125, 127, 129,
        131, 167
Stuxnet 5–6, 43
substitution 60, 108–111, 124
suspects 35, 37–40, 44
Sweeney, LaTanya 15, 26, 56, 58,
        160–161, 164, 166
symmetric encryption 78, 119
systemizing 31

theft 2, 9, 33, 41–43
theoretical perspective 24–25, 83
threats 12, 36–37, 40, 126, 166

token 46, 49, 51, 76
transmission 59, 107, 126
transposition 60–61, 67, 108, 111
Turing, Alan 14, 66, 86–91, 95–96,
        167
Tversky, Amos 139–141, 146–147
two-factor authorization 49
two-person zero-sum game 69,
        72, 74

United States Presidential Election
        6–11, 29, 36, 56, 108,
        148, 153, 160, 164, 166
Universidad Santo Tomas 4
University of Buea 132
UNIX 1, 52
utility theory 72, 140, 142, 165

Venn diagram 54, 57
Vigenère cipher 62, 66–67
Vigfusson, Ymir 32–34
virus 9–10, 154
Von Neumann, Janos 69, 72, 79
vulnerabilities 1, 32

WannaCry 7, 11, 165
watermarking, digital 124
Watson 90, 95–96
Welles, Orson 148
Wiener, Anna 33, 34
WikiLeaks 2–3, 7, 10–11, 39, 54,
        77–78
World Infrastructure Security
        Report 4, 11, 15, 66, 88,
        90, 96, 125–126, 159
worms 9

Yahoo 148, 151
Yu, Pingan (Gold Sign) 8

Zombie Apocalypse 154